2nd International Conference on Energy Materials and Applications (ICEMA 2017)

IOP Conference Series: Materials Science and Engineering
Volume 222

Hiroshima, Japan
10 – 12 May 2017

ISBN: 978-1-5108-4512-1
ISSN: 1757-8981

Printed from e-media with permission by:

Curran Associates, Inc.
57 Morehouse Lane
Red Hook, NY 12571

Printed by Curran Associates, Inc. (2017)

For permission requests, please contact the Institute of Physics
at the address below.

Institute of Physics
Dirac House, Temple Back
Bristol BS1 6BE UK

Phone: 44 1 17 929 7481
Fax: 44 1 17 920 0979

techtracking@iop.org

Additional copies of this publication are available from:

Curran Associates, Inc.
57 Morehouse Lane
Red Hook, NY 12571 USA
Phone: 845-758-0400
Fax: 845-758-2633
Email: curran@proceedings.com
Web: www.proceedings.com

Table of contents

Volume 222

2nd International Conference on Energy Materials and Applications
10–12 May 2017, Hiroshima, Japan

Accepted papers received: 10 July 2017
Published online: 24 July 2017

Key technologies and applications of functional materials

Smart grid and electrical automation

IOP Conf. Series: Materials Science and Engineering **222** (2017) 011001 doi:10.1088/1757-899X/222/1/011001

PREFACE

It is our great pleasure to welcome you to 2017 2nd International Conference on Energy Materials and Applications which will take place in Hiroshima, Japan during May 10-12, 2017. ICEMA2017 is dedicated to issues related to Energy Materials and Applications.

The major goal and feature of the conference is to bring academic scientists, engineers, industry researchers together to exchange and share their experiences and research results, and discuss the practical challenges encountered and the solutions adopted. Professors from Singapore, Taiwan and Japan are invited to deliver keynote speeches regarding latest information in their respective expertise areas. It will be a golden opportunity for the students, researchers and engineers to interact with the experts and specialists to get their advice or consultation on technical matters, sales and marketing strategies.

These proceedings present a selection from papers submitted to the conference from universities, research institutes and industries. All of the papers were subjected to peer-review by conference committee members and international reviewers. The papers selected depended on their quality and their relevancy to the conference. The volume tends to present to the readers the recent advances in the field of Energy Materials and various related areas, such as performance evaluation and testing of building materials, key technologies and applications of functional materials, smart grid and electrical automation, etc.

We would like to thank all the authors who have contributed to this volume and also to the organizing committee, reviewers, speakers, chairpersons, sponsors and all the conference participants for their support to ICEMA 2017.

Prof. Dr. Lu Li
National University of Singapore, Singapore
May 20, 2017

IOP Conf. Series: Materials Science and Engineering **222** (2017) 011002 doi:10.1088/1757-899X/222/1/011002

COMMITTEES

Advisory Chair

Prof. Yanglong Hou, Peking University, China
Prof. Toshifumi Ise, Osaka University, Japan

Conference Chairs

Prof. David Lin, National University of Tainan, Taiwan
Prof. LU Li, National University of Singapore, Singapore

Technical Program Committee

Assoc. Prof. Dr Yamuna a/p Munusamy, Universiti Tunku Abdul Rahman, Malaysia
Assoc. Prof. Dr. Weiwei Cai, China University of Geosciences Wuhan, China
Assoc. Prof. Dr. Noor Asmawati Mohd Zabidi, Universiti Teknologi PETRONAS, Malaysia
Dr. Marwan Marwan, Syiah Kuala University, Indonesia
Dr. Frank Römer, Univeristy of Bonn, Germany
Dr. Lei Liu, Univeristy of Bonn, Germany
Dr. S. Harish, Kyushu University, Japan
Dr. Geetesh Goga, K.C. College of Engineering & IT, India
Prof. Koteeswara Reddy Nandanapalli, Sungkyunkwan University, South Korea
Assoc. Prof. Yubao Sun, China University of Geosciences (Wuhan), China
Assoc. Prof. Zhang Yunfeng, China University of Geosciences (Wuhan), China
Asst. Prof. MILIND SURYAJI PATIL, Guru Gobind Singh College, INDIA
Dr. Parnuwat Usapein, Rajamangala University of Technology Rattanakosin, Thailand
Asst. Prof. Arpit Jain, University of Petroleum and Energy Studies, India
Dr. Okoye Celestine Nwoye, Federal Ministry of Education, Nigeria
Dr. Om Prakash Yadav, Haramaya University, Ethiopia
Dr. Ramadhansyah Putra Jaya, Universiti Teknologi Malaysia, Malaysia
Dr. Uthen Thubsuang, Walailak University, Thailand
Dr. Weerapol Namboonruang, Muban Chombueng Rajabhat University, Thailand
Prof. G. Selvakumar, Muthayammal Engineering College, India
Prof. Ya-Fen Wu, Ming Chi University of Technology, Taiwan
Dr. Sumate Lipirodjanapong, Muban Chombueng Rajabhat University, Thailand
Dr. Nantamol Limphitakphong, Chulalongkorn University, Thailand
Dr. Aman Mahajan, Guru Nanak Dev University, India
Dr. Wen-Jay Lee, National Center for High-Performance Computing, Taiwan
Prof. Jesus Toribio, University of Salamanca, Spain
Asst. Prof. Diego Vergara, University of Salamanca, Spain
Asst. Prof. Miguel Lorenzo, University of Salamanca, Spain

Peer review statement

All papers published in this volume of *IOP Conference Series: Materials Science and Engineering* have been peer reviewed through processes administered by the proceedings Editors. Reviews were conducted by expert referees to the professional and scientific standards expected of a proceedings journal published by IOP Publishing.

IOP Conf. Series: Materials Science and Engineering **222** (2017) 012001 doi:10.1088/1757-899X/222/1/012001

Voids characteristics of asphaltic concrete containing coconut shell

Mohd Ezree Abdullah[1], Amirah Hannani Madzaili[2], Ramadhansyah Putra Jaya[2,3] Haryati Yaacob[2], Norhidayah Abdul Hassan[2], Fadzli Mohamed Nazri[4]

[1]Faculty of Civil and Environmental Engineering, Universiti Tun Hussein Onn Malaysia, 86400 Batu Pahat, Johor, Malaysia
[2]Faculty of Civil Engineering, Department of Geotechnics and Transportation, Universiti Teknologi Malaysia, 81310 Johor Bahru, Malaysia
[3]Faculty of Engineering, Department of Civil and Environmental Engineering, Universiti Teknologi PETRONAS, 32610 Seri Iskandar, Perak Darul Ridzuan, Malaysia
[4]School of Civil Engineering, Universiti Sains Malaysia, Engineering Campus,14300 Nibong Tebal, Penang, Malaysia

Email: ramadhansyah@utm.my

Abstract. Asphalt durability is often linked to the thickness of the asphalt coating on the aggregate particles. In order to have adequate film thickness in asphaltic concrete, there must be sufficient space between the aggregate particles in the compacted pavement. This void space is referred to as voids in total mix (VTM), voids with filled bitumen (VFB), and voids in mineral aggregate (VMA). Hence, this study investigates the performance of coconut shell (CS) as coarse aggregate replacement on voids characteristics of asphaltic concrete. Four CS were used as coarse aggregates replacement in asphalt mixture namely 0%, 10%, 20%, 30%, and 40% (by weight volume). The voids properties of asphalt mixture were determined based on Marshall Mix design test. Test results show that VTM and VMA values were decrease with the increasing bitumen content where VFB was increase with increasing bitumen content. Furthermore, increasing the percentage of coconut shell in asphalt mixture was found to increases the voids value up to a peak level and then decreases with further additions of CS.

1. Introduction

Conventional aggregates sources in Malaysia are depleting fast due to speedy growth of construction activities [1]. Currently, transportation industry is under pressure to use alternative materials due to high consumption of aggregates in road construction [2]. In order to move towards sustainable development, these materials should be used wisely. A large number of researches have been done in order to search for a possible material as aggregates replacement in road conductions [3,4,5]. Currently, using agricultural waste materials as an aggregates replacement becomes trends in construction industry [6,7]. The aim of using agricultural waste is to minimize the production of waste, reserve the environment and improve the pavement durability [8]. Durability refers to the ability of the mixture to resist abrasion of the surface due to the voids content is high allowing air and water to prematurely harden the asphalt. Coconut shell is one of the agricultural wastes that can be used to durability resistance [9]. Several of research has been done by using coconut shell as coarse aggregates in lightweight concrete [10,11]. Yet, the usage of coconut shell in hot mix asphalt has never been extensively studied. Therefore, this study presents the effects of coconut shell on the voids characteristics of hot mix asphalt (HMA). This investigation focused on using the variation of coconut shell from 0% to 40% as aggregate replacement in the HMA.

2. Materials and methods

2.1. Binder
The bitumen used in this study was a 60/70 PEN supplied by Chevron Malaysia. This grade bitumen is widely used in Malaysia and other countries. According to the manufacturer's specification, the relative density value of the 60/70 binder was 1.03. The softening point and penetration was 52°C and 65dmm, respectively.

2.2. Aggregate and gradation
Granite aggregates supplied by Kuad Quarry Sdn. Bhd. was used throughout this investigation. Prior to aggregate batching, the aggregates were washed, dried and sieved into their respective size ranges. The granite aggregate had a specific gravity of 2.65 g/cm^3, while the ACV and water absorption was 26.1% and 0.48%, respectively. In addition, the aggregate gradation used in this study conformed to JKR/SPJ/2008-S4 [12] specifications for asphaltic concrete AC14.

2.3. Mix preparation
Ordinary Portland cement and hydrated lime were used as filler in this investigation. The 60/70 PEN bitumen was used as binder for mix preparation. The aggregates were first mixed into batches according to the designated gradations and weight. These batches were then heated in an oven at the designated mixing temperature for at least 4 hours before the mixing process. The heated aggregate batches were then mixed with a specified amount of bitumen. The mixes were compacted with 75 blows on each side with the standard Marshall hammer. After compaction, the specimens were removed from the molds and allowed to cool overnight.

2.4. Voids characteristics
The air voids characteristics of asphaltic concrete AC14 was determined based on Marshall Mix design as outlined in ASTM D6927-15 [13]. The procedure aimed at determining a number of parameters including voids in total mix, voids filled with bitumen, and voids in mineral aggregate. For this purpose, the specimens were prepared at binder contents ranging from 4.0 to 6.0% by weight in 1.0% increment.

3. Results and discussion

3.1. Voids in total mix
VTM is the total volume of the small pockets of air between the coated aggregate particles throughout a compacted paving mixture, expressed as a percent of the compacted mixture. The effects of asphaltic concrete incorporating coconut shell on VTM are presented in figure 1. Generally, voids in total mix decrease gradually with the increasing of bitumen content. For instance, at 0% coconut shell mix, the VTM value of 8.5 to 2.8% was achieved when the bitumen content was increased from 4.0 to 6.0%. On the other hand, increasing the percentage of coconut shell in asphalt mixture was found to increases the VTM values up to a peak level and then decreases with further additions of CS. For example, the VTM value of 8.5 to 13.6% was achieved at 4% bitumen content when the coconut shell level was increased from 0 to 40%. Based on JKR specification [12], the VTM for wearing course should range from 3.0% to 5.0%. However, the requirement does not specify limitation values for asphalt mixture incorporated waste agriculture materials, especially coconut shell. At all VTM investigated, the 0% CS content mixes record the lowest voids in total mix. Hassan et al. [14] reported

that lower air void contents minimize the aging of the asphalt cement films within the aggregate mass and also minimize the possibility of moisture penetrating the thin asphalt cement film and strip the asphalt cement off aggregates. In this investigation, the lowest VTM value of asphaltic concrete incorporating coconut shell was recorded at 10% and 40%, respectively.

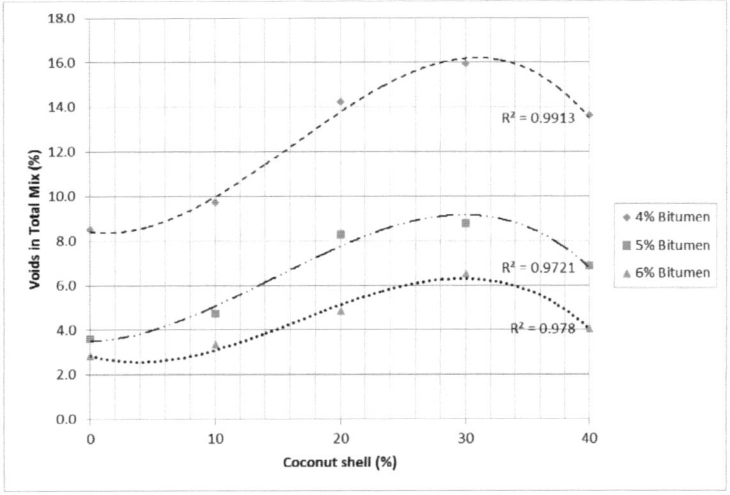

Figure 1. VTM vs. coconut shell at varying bitumen content

3.2. Voids in mineral aggregates

The VMA values of asphaltic concrete at varying percentage coconut shell are illustrated in figure 2. It can be seen that the VMA values of specimens are decrease by increasing bitumen content. For instance, at 20% CS, the VMA value of 20 to 14% was achieved when the bitumen content was increased from 4.0 to 6.0%. The CS had a lower specific gravity due to lightweight material but higher water absorption, indicating that it absorbs more bitumen content during the mixing process. Similar findings were also reported by another investigation [7,9,15]. The results also indicated that the VMA of controlled specimen is lowest than VMA of asphaltic concrete containing 10, 20, 30, 40% coconut shell, respectively. There are no specifying limitation values for VMA in JKR standard for asphaltic concrete containing waste materials. The VMA value was found to range from 13 to 19% when the coconut shell content was increased from 0 to 40%.

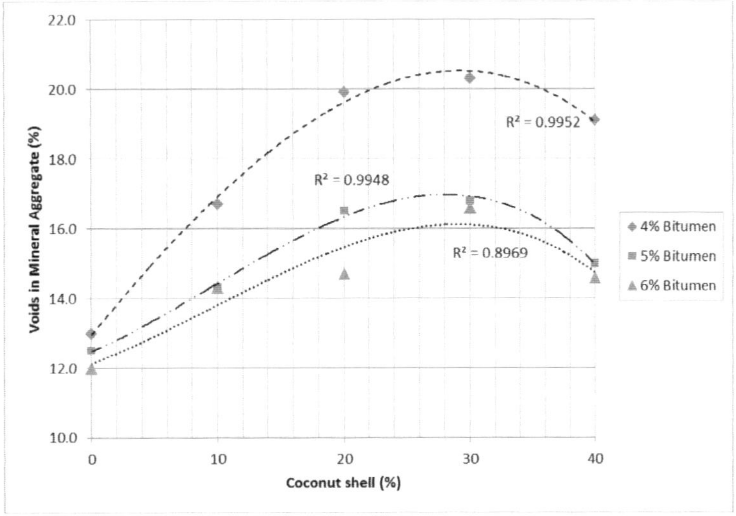

Figure 2. VMA vs. coconut shell at varying bitumen content

3.3. Voids filled with bitumen

Voids filled bitumen (VFB) is the percentage of the volume of the VMA that is filled with bitumen. The VFB is inversely related to the air voids. As the percentage of air voids approaches zero, the VFB approaches 100 [7]. The general effect of asphaltic concrete containing coconut shell on VFB at different bitumen content is illustrated in figure 3. It can be seen that the VFB decreases as the coconut shell quantity increases. The VFB reduces from 54% to 44% as the coconut shell content increases from 0% to 40%. JKR specification [16] stated that the VFB value of specimen should range from 75% to 85%. However, these specifications do not specify limiting values for agriculture waste materials, specifically for coconut shell. The results also show that the VFB values of specimens are increase by increasing bitumen content. For instance, at 0% CS, the VFB value of 54 to 79% was achieved when the bitumen content was increased from 4.0 to 6.0%. Compared with the study conducted by Jeffry et al. [12] reported that when the VFB exceeds approximately 80% to 85%, the asphaltic concrete mix typically becomes unstable and rutting is likely to occur. Hence, the VFB values are within 39% to 54%, 65% to 80%, and 79% to 97%, correspondingly for 4%, 5%, and 6% bitumen content. It can be concluded that the use of 0% to 40% CS in asphalt mixture at 4% bitumen content was more effective in enhancing rutting.

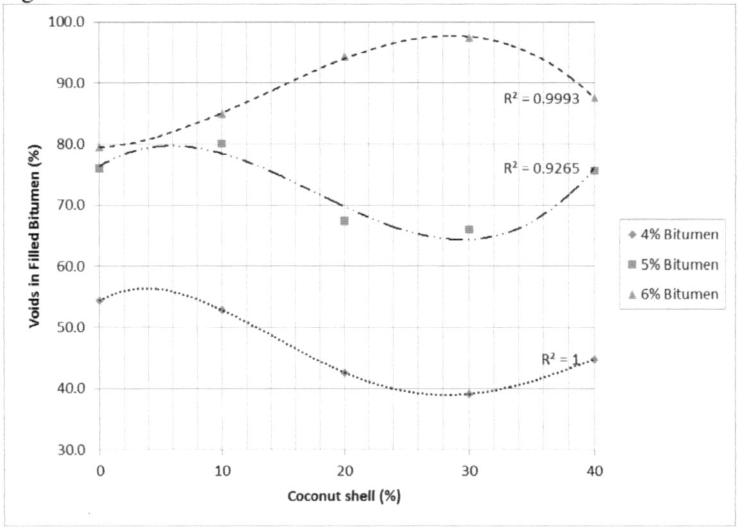

Figure 3. VFB vs. coconut shell at varying bitumen content

4. Conclusions

The effect of coconut shell on the voids properties of asphalt mixture was investigated. It can be seen that the addition of coconut shell as aggregate replacement materials was not enough improvement to the durability performance of asphaltic concrete. However, when compared among substitutes level, the use of 10% CS resulted in good resistance of mixtures.

References

[1] Jusli E, Nor H M, Jaya R P and Haron Z, Strength and microstructure properties of double layered concrete paving blocks containing waste tyre rubber granules. *Jurnal Teknologi* Vol. 73 (2015), p. 85.

[2] Abdullah M E, Zamhari K A, Hainin M R, Oluwasola E A, Yusoff N I Md. and Hassan N A, High temperature characteristics of warm mix asphalt mixtures with nanoclay and chemical warm mix asphalt modified binders. *Journal of Cleaner Production* Vol. 122 (2016), p. 326.

[3] Hidayah N A H, Hasanan M and Ramadhansyah P J, Effect of coarse aggregate sizes on properties of Porous Concrete Paving Blocks. *Advanced Materials Research* Vol. 911 (2014), p. 433.

[4] Abdullah M E, Zamhari K A, Buhari R, Kamaruddin N H M, Nayan N, Hainin M R, Hassan N A, Jaya R P and A Yusoff N I M: review on the exploration of nanomaterials application in pavement engineering, *Jurnal Teknologi* Vol. 73 (2015), p. 69.

[5] Tarmuzi N A, Jaya R P, Yaacob H, Hassan N A and Aziz M M A: Aggregate angularity effect on porous asphalt engineering properties and performance, Jurnal Teknologi, Vol. 73 (2015), p. 99.

[6] Buhari R, Zabidi N F, Abdullah M E , Bakar S K A and Kamarudin N H M: Rheological behaviour of coconut shell powder modified asphalt binder, Jurnal Teknologi Vol. 78 (2016), p. 17.

[7] Norhafizah M, Ramadhansyah P J, Siti Nur Amiera J, Nurfatin Aqeela M, Norhidayah A H, Hainin M R and Che Norazman C W: The effect of coconut shell on engineering properties of porous asphalt mixture. *Jurnal Teknologi* Vol. 78 (2016), p. 127-132.

[8] Ma B, Adhikari S, Chang Y, Ren J, Liu J and You Z: Preparation of composite shape-stabilized phase change materials for highway pavements. *Construction and Building Materials* Vol. 42 (2013), p. 114.

[9] Ramadhansyah P J, Nurfatin Aqeela M, Siti Nur Amiera J, Norhafizah M, Norhidayah A H and Dewi S J: Use of coconut shell from agriculture waste as fine aggregate in asphaltic concrete. ARPN *Journal of Engineering and Applied Sciences* Vol. 11 (2016), p. 7457.

[10] Ali M, Liu A, H. Sou and Chouw N: Mechanical and dynamic properties of coconut fibre reinforced concrete. Construction and Building Materials Vol. 30 (2012), p. 814.

[11] Prusty J K and Patro S K: Properties of fresh and hardened concrete using agro-waste as partial replacement of coarse aggregate – A review. *Construction and Building Materials* Vol. 82 (2015), p. 101.

[12] Jeffry S N A, Jaya R P, Manap N, Miron N A and Hassan N A: The influence of coconut shell as coarse aggregates in asphalt mixture. *Key Engineering Materials* Vol. 700 (2016), p. 227.

[13] ASTM D6927-2015. Standard Test Method for Marshall Stability and Flow of Bituminous Mixtures. *Annual Book of ASTM Standards*, West Conshohocken, Philadelphia, U.S.A.

[14] Hassan N A, Mahmud M and Jaya R P, Air void characterisation in porous asphalt using x-ray computed tomography. *Advanced Materials Research* Vol. 911 (2014), p. 443.

[15] Ting T L, Jaya R P, Hassan N A, Yaacob H, Jayanti D S and Ariffin M A M: A review of chemical and physical properties of coconut shell in asphalt mixture. *Jurnal Teknologi* Vol. 78 (2016), p. 85.

[16] JKR/SPJ/2008-S4. 2008. *Standard Specification for Road Works*. Malaysian Public Works Department.

Acknowledgements

Research work presented in this paper was supported by Ministry of Higher Education Malaysia (MOHE), Fundamental Research Grant Scheme (grant number 1458) and Incentive Grant Scheme for Publication (grant number U669), Universiti Tun Hussein Onn Malaysia; and Universiti Teknologi Malaysia (grant number Q.J130000.2522.11H58). The financial assistance for this study is highly appreciated.

IOP Conf. Series: Materials Science and Engineering **222** (2017) 012002 doi:10.1088/1757-899X/222/1/012002

Techno – economic and leachability effect of Ca2SiO4 interlocking composite brick

Weerapol Namboonruang[1 a], Prayoon Yongam-nuai[1 b] and Nutthanun Suphadon[2 c]

[1]Department of Civil Technology and Architecture, Faculty of Industrial Technology, Muban Chombueng Rajabhat University, MCRU 70150, Thailand
[2]National Metal and Materials Technology Center 12120, Thailand

Email: [a]bannork_civil@hotmail.com , [b]ajanyoon@hotmail.com , [c]nutthans@mtec.or.th

Abstract. This work studied the possibility to produce the new novel calcium silicate ceramic kiln ash composite brick (CSCACB) which was the combination of the local Ratchaburi soil and calcium silicate kiln ash (CSCA) wasted from the ceramic industry. The chemical and physical properties of the composite brick were investigated. Also, the mechanical properties such as the compressive strength, flexural strength and as well as the leachability property were determined. By summary, this invented materials has enough quality to produce as local commercial products considered by the strength properties, environmental effect, price and appropriated implementation.

1. Introduction

Nowadays, the trend of global warming campaign is prevalent in many countries, including Thailand, which was ranked 71st in the world in gas emissions that destroy the Earth's atmosphere. There are many reasons causes of the global warming, for instance emissions from vehicles, factories, residences etc. To reduce these effects, green product or productions are currently used for example, the ecology house and so on. As mentioned above, over the past decade, the local manufacturing sector is one of many industries that use the thermal energy as the main source in the production process as well as a large factory, but are only different in the type of fuel for energy purposes. By collecting data, it was found that number of the local ceramic industrial in Ratchaburi province are 42 industries which 85%are a semi-ancient [1]. This is to say that the traditional ceramic kiln use the fuel supplied from thermal products of wood such as wood fragmented and small trees.

Resulting from this process is the environmental issues that arise while burning such as fumes from fuel and ash from the ceramic kiln and the construction industry uses ash in the products and to replace cement in concrete [2], for soil stabilization and in cement production [3-8]. Therefore, we have adopted this type of ash to study and analyze the features design as a soil brick with bearing and non-bearing types [9,10]. The purpose is to develop and design the local materials using for a low-energy building construction considered by the suitability of the production cost, standard, strength, appropriate technology and being environmentally friendly building materials that can be utilized for local residents as well as technology and research to continue in small and medium-sized communities effectively.

2. Methodology

For the property testing of soil bricks, samples were molded approximately 1,085 samples with the dimension of 125 mm in width, 500 mm in length and 100 mm in thickness. Types of brick sample were classified below, Unreinforced Soil Cement Composite Brick (filled mortar-USCCB) (control). This type of brick was filled with mortar 1:3 when was constructed as the composite system. Unreinforced Soil Cement Composite Brick (unfilled mortar-USCCB). Calcium Silicate Ceramic Kiln Ash Composite Brick(CSCACB). This type of brick was filled with mortar 1:3 when was constructed as the composite system. Calcium Silicate Ceramic Kiln Ash Composite Brick (unfilled mortar-CSCACB). For bricks type a and b, the cement contents were varied at *8*, 13, 18, 23, 28, 33, 38, 43, 48 and *53* % by weight of soil basis at the water content of *9.74*, 10.29, 10.52, 10.75, 10.88, 11.01, 11.38, 11.74, 13.31 and *14.87* at cure time of *3*, 7, 14, 28, 60, 90, 120, 150 and *180* days, respectively. For bricks type c and d, a constant Portland cement type I content (OPC) of 3% was mixed with varying calcium silicate kiln ash (CSCA) contents of 15, 20, 25, 30, 35, 40, 45 and 50 % at cure time of 7, 14, 28, 60, 90, 120, 150 and 18*0* days, respectively.

3. Results and discussions

3.1. Chemical properties

Table 1 shows the chemical property of three material types such as Non-plastic Soil (NP-S), Ordinary Portland cement type I (OPC) and Calcium silicate ceramic kiln ash (CSCA) by using X-Ray Fluorescence (XRF). The total of the main oxide of calcium silicate ceramic kiln ash which effects to the strength development reaction such as SiO_2, CaO and Al_2O_3 is 48.32% and 90.31% for the ordinary Portland cement type I (OPC) as well as 80.25% for the local Ratchaburi soil. Also, the loss of ignition (LOI) of materials is 29.12, 2.31 and 2.19% respectively.

Table 1. The chemical composition of non-plastic Soil (NP-S), ordinary Portland cement (OPC) and calcium silicate ceramic kiln ash (CSCA) using X-ray fluorescence (XRF).

Oxides	Concentrations		
	Non-plastic Soil (NP-S)	Ordinary Portland cement type I (OPC)	Calcium silicate ceramic kiln ash (CSCA)
Al_2O_3	24.73	4.77	1.26
SiO_2	37.89	19.83	4.31
SO_3	5.88	2.05	1.21
CaO	17.63	65.71	42.75
Fe_2O_3	7.85	3.82	0.20
LOI	2.19	2.31	29.12

3.2. Mechanical properties

3.2.1. Compressive strength of unreinforced Soil Cement Composite Brick (filled and unfilled mortar-USCCB). Figure 1(a) and 1(b) show the compressive strength of both filled and unfilled mortar of USCCB with varying cement contents of 8, 13, 18, 23, 28, 33, 38, 43, 48 and 53% by weight of soil basis at the cure time of *3*, 7, 14, 28, 60, 90, 120, 150 and *180* days with water content of *9.74*, 10.29, 10.52, 10.75, 10.88, 11.01, 11.38, 11.74, 13.31 and *14.87*% respectively. Both graphs show a similar trend which the compressive strength increases with increasing the cement contents and cure time. However, the increasing rate is not significant when the cure time is longer than 60 days. Comparing between filled and unfilled mortar, at cure time of 28 days the maximum compressive strength of filled and unfilled mortar occurs at the same cement content of 53% with values of *221.22* $kg.cm^{-2}$and 121.25$kg.cm^{-2}$ respectively. Results show that the compressive strength of unfilled mortar is significant lower than those of filled mortar-USCCB. This can be explained that filled mortar USCCB has a larger cross-section area which is proportional to the compressive strength and the strength of the mortar is relatively high compared to USCCB.

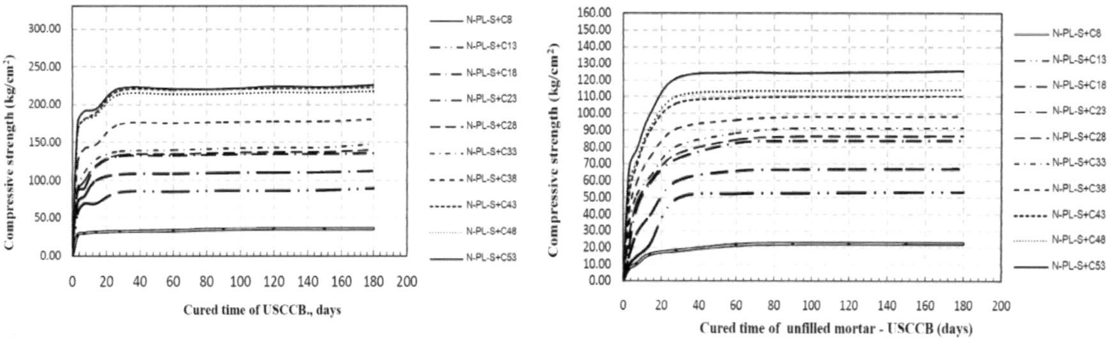

(a) (b)

Figure 1. The compressive strength of unreinforced soil cement composite brick for filled mortar USCCB (a). and unfilled mortar-USCCB (b).

Figure 2. The unreinforced soil cement composite brick for filled mortar USCCB and unfilled mortar-USCCB (size 125x500x100 mm3)

3.2.2. Compressive strength of calcium silicate ceramic kiln ash composite brick (filled and unfilled mortar CSCACB). Figure 3(c) and 3(d) show the compressive strength of filled and unfilled mortar with varying the calcium silicate ceramic kiln ash content of *5*, 10, 15, 20, 25, 30, 35, 40, 45 and *50* at the cure time of *3*, 7, 14, 28, 60, 90, 120, 150 and *180* days. Results reveal that the compressive strength decreases with increasing of calcium silicate ceramic kiln ash. The maximum compressive strength occurs at calcium silicate ceramic kiln ash content of 5% with values of *39.01* kg.cm^{-2}and 27.71 kg.cm^{-2} for filled and unfilled mortar CSCACB, respectively at cured time of 28 days.

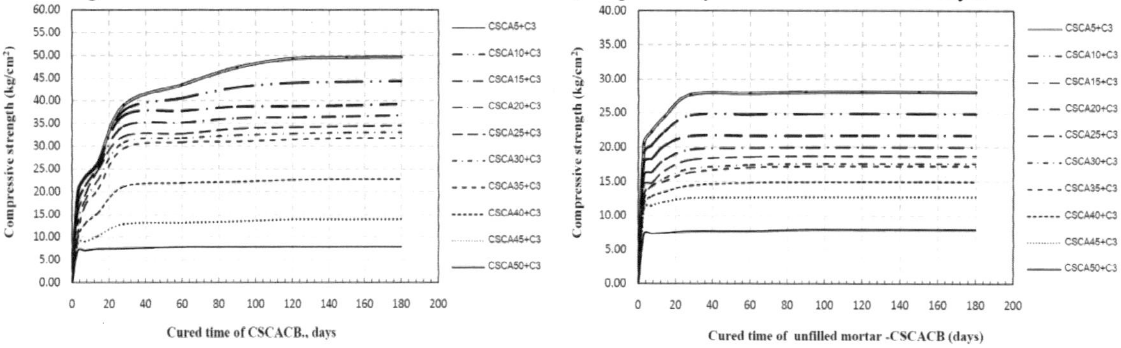

(c) (d)

Figure 3. The compressive strength of Calcium Silicate Ceramic Kiln Ash Composite Brick for filled mortar –CSCACB (c). and unfilled mortar –CSCACB (d).

*3.2.3. Flexural strength on Modulus of rupture of Unreinforced Soil Cement Composite Brick(filled and unfilled –mortar USCCB)*The Flexural strength of both filled and unfilled mortar USCCB is presented in Figure 4(e) and 4(f) respectively. Agreement to the compressive strength, the flexural strength increases with increasing of cement content for both cases. At 28 days, the maximum value of filled and unfilled mortar USCCB are showed at the cement content of 53% which values are *15.176* kg.cm^{-2} and *8.063* kg.cm^{-2}. Form these results, it can be seen that the flexural strength of filled is higher than those of unfilled mortar approximately as 53.13%.

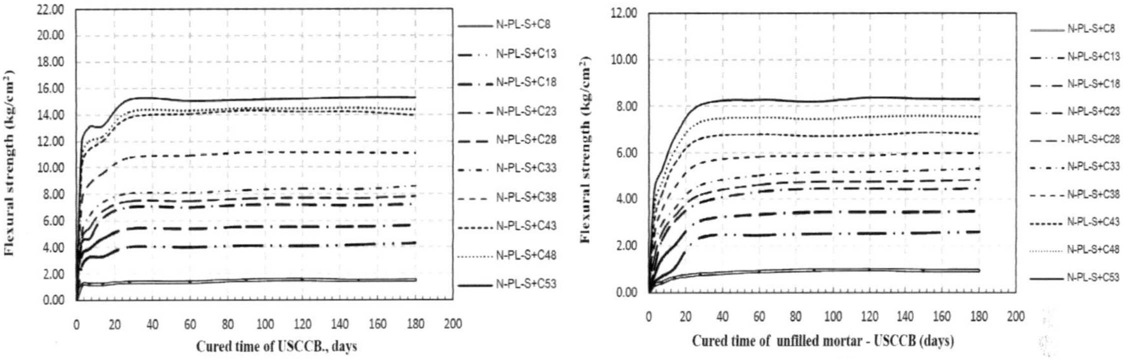

(e) (f)

Figure 4. The flexural strength of unreinforced soil cement composite brick for filled mortar USCCB (e). and unfilled mortar-USCCB (f).

3.2.4. Flexural strength on modulus of rupture of calcium silicate ceramic kiln ash composite brick (filled and unfilled mortar-CSCACB) Similarly to the compressive strength, flexural strength on modulus of rupture of calcium silicate ceramic kiln ash composite Brick(filled and unfilled mortar-CSCACB) decrease with increasing the contents of calcium silicate ceramic kiln ash as shown in Figure 5(g) and 5(h). At 28 days, the maximum value of filled and unfilled mortar CSCACB are *1.62* kg.cm^{-2} and *1.14* kg.cm^{-2} occurred at 5% calcium silicate ceramic kiln ash. Moreover if compare to previous results of USCCB at 53% cement content for both filled and unfilled mortar, CSCACB5+3 are lower than those of USCCB at 53% considered as 10.67% and 14.13% respectively. According to the strength properties of materials, it can be concluded that:

The ratio of the flexural strength in rupture to the compressive strength of USCCB (filled mortar) and CSCACB (filled mortar) is approximated as *0.96 ~ 1.00*. The ratio of the flexural strength in rupture to the compressive strength of USCCB (unfilled mortar) and CSCACB (unfilled mortar) is approximated as ~ *1.00*. When considers the strength of filled and unfilled mortar CSCACB, it is clearly seen than none of them can qualify the bearing load type standard according to Thai Industrial Standards Institute 102-2517 [9] and the community standard 602-2547 [10]. However, with calcium silicate ceramic kiln ash of 5%and cement of 3% (CSCACB5+3) can be still used as the non-bearing type product.

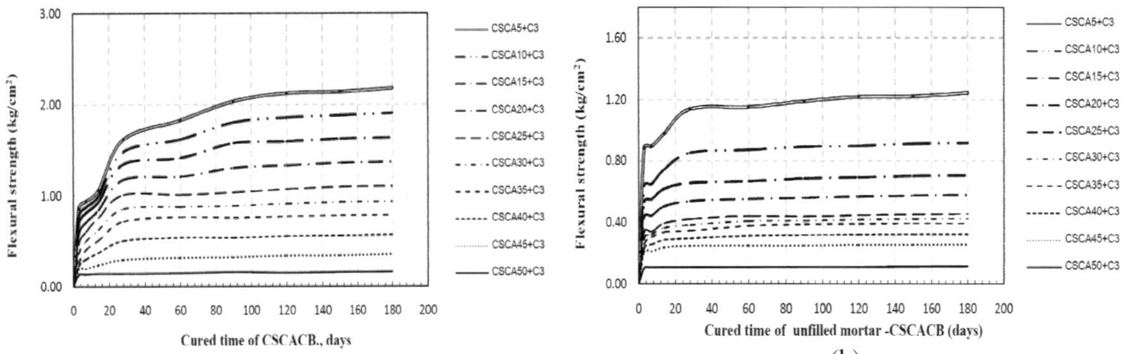

(g) (h)

Figure 5. The flexural strength on modulus of rupture of calcium silicate ceramic kiln ash composite brick for filled mortar-CSCACB (g). and unfilled mortar-CSCACB (h).

3.3. An economically assessment of products

The cost analysis of the building prototype constructed by CSCACB5+C3 is compared to other brick types such as cement brick, light weight brick, composite brick and local soil brick. The price of CSCACB5+C3 at the total water ratio of 12.69% is calculated as 3.6 baht/piece. For the price of other bricks used for construction with the non-bearing wall type in an area of Chombueng district, Ratchaburi during 2015-2016 is given as 4.5 baht/piece for cement brick, 0.8 baht/piece for local soil brick, 8 baht/piece for composite brick and 18 baht/piece for light weight brick. As this information, it can be seen that the price of CSCACB5+C3 is lower than other brick except the local soil brick. The cost summary of the wall constructed by each brick type is shown in table 2.

Table 2. The price of brick products in area of Chombueng district studied during 2015-2016.

Product	Product price(Baht/m^2)	Mortar 1 :3 (Baht/m^2)	Labor price(Baht/m^2)	Plaster price (Baht/m^2)	Total price (Baht/ m^2)
CSCACB5+C3**	144	64	80	None	288
Lightweight brick	162	50	80	190	484
Composite brick	320	64	80	None	464
Local soil brick	140	64	80	190	474
Cement brick	55	50	80	190	375

** CSCACB5+C3 is appropriated proportion of Ca_2SiO_4 interlocking composite brick

3.4. Leachate test

Accord to the study above, it found that the suitable formulation to produce as the commercial brick is CSCACB with 3% cement content and 5% calcium silicate ash by dry soil weight, CSCACB5+C3 (an appropriated proportion of Ca_2SiO_4 interlocking composite brick). Then this brick was tested to find the concentration of 8 types of hazardous substances following by U.S EPA SW-846(1992) Method 1311 [11]. Result shows that the concentrations of hazardous substances are lower than the maximum standard limit as shown in table 3. For environmental effect and the safety of the residents, with considered by leachate test, it is found that there are virtually no heavy metals and deterioration of hazardous substances such as Lead (Pb). However, only Chromium (Cr) and Arsenic (As) are found with small amount of quantity compared to the standard limit of 5.00 mg.l^{-1}.

Table 3. The leachate test of CSCACB5+C3 (an appropriated proportion of Ca2SiO4 interlocking composite brick)

Metals	CSCACB3+5**	Standard
Arsenic (As)	0.15	not more than 5.0 mg.l^{-1}
Barium (Be)	None	not more than 100.0 mg.l^{-1}
Cadmium (Cd)	None	not more than 1.0 mg.l^{-1}
Chromium(Cr)	0.10	not more than 5.0 mg.l^{-1}
Lead (Pb)	None	not more than 5.0 mg.l^{-1}
Mercury (Hg)	None	not more than 0.20 mg.l^{-1}
Selenium (Se)	None	not more than 1.0 mg.l^{-1}
Silver (Ag)	None	not more than 5.0 mg.l^{-1}

** CSCACB5+C3 is appropriated proportion of Ca_2SiO_4 interlocking composite brick

4. Conclusions

According to this study, varying the contents of calcium silicate ceramic kiln ash was added in soil cement brick. The purpose is to apply the use of waste from the ceramic industry to develop as the local commercial product. Results show that both compressive strength and flexural strength decrease with increasing calcium silicate ceramic kiln ash content. At cured time of 28 days, it is found that the maximum compressive strength occurs at cement content of 3% and calcium silicate ceramic kiln ash of 5% (CSCACB5+3) with values of *39.01* kg.cm^{-2}(filled mortar) and *27.71* kg.cm^{-2} (unfilled mortar) respectively. Also, the maximum flexural strength occurs at the same content(CSCACB5+3) which values are *169* kg.cm^{-2} (filled mortar) and *1.639*kg.cm^{-2} (unfilled mortar) respectively.

By comparing between filled and unfilled mortar of materials in this work, results reveal that filling mortar in the hole of the brick can increase the strength of the brick composite unsurprisingly. That means CSCACB is a better heat resistant material compared to USCCB. For environmental factor, result shows that the product pass the hazardous substances testing following by U.S EPA SW-846 Method 1311(TCPL). Finally, according to the property investigation in this work, it can be concluded that CSCACB5+C3(Ca_2SiO_4 interlocking composite brick) shows an enough quality to produce as the local economic material effectively considered by various factors such as the strength properties, heat resistance, low price, simple technology, and safety including environmental friendly.

References

[1] W. Namboonruang, R. Rawangkul, W. Yodsudjai, T. Aramraks and N. Suphadon, Prediction of the Heat-Insulating Crumb Rubber Brick Walls Design by the Finite Element Method. Adv.Mater.Res. 805-806(2013), pp.1575-1582.

[2] J. Cuenca, J. Rodríguez, M. Martín-Morales, Z. Sánchez-Roldán, M. Zamorano, Effects of olive residue biomass fly ash as filler in self-compacting concrete, Constr. Build. Mater. 40 (2013) 702–709.

[3] S.V. Vassilev, D. Baxter, L.K. Andersen, C.G. Vassileva, An overview of the composition and application of biomass ash. Part 2. potencial utilization, technological and ecological advantages and challenges, Fuel 105 (2013) 19–39.

[4] R. del Valle-Zermeño, J.M. Chimenos, J. Giró-Paloma, J. Formosa, Use of weathered and fresh bottom ash mix layers as a subbase in road constructions: environmental behavior enhancement by means of a retaining barrier, Chemosphere 117 (2014) 402–409.

[5] T. Hemalatha, Maitri Mapa, Neenu George, Saptarshi Sasmal, Physico-chemical and mechanical characterization of high volume fly ash incorporated and engineered cement

IOP Conf. Series: Materials Science and Engineering **222** (2017) 012002 doi:10.1088/1757-899X/222/1/012002

system towards developing greener cement, J. Clean. Prod. 125 (2016) 268–281.
[6] M. Rafieizonooz, J. Mirza, M. Razman Salim, M. Warid Hussin, E. Khankhaje, Investigation of coal bottom ash and fly ash in concrete as replacement for sand and cement, Constr. Build. Mater. 116 (2016) 15–24.
[7] I. Garcia-Lodeiro, V. Carcelen-Taboada, A. Fernández-Jiménez, A. Palomo, Manufacture of hybrid cements with fly ash and bottom ash from a municipal solid waste incinerator, Constr. Build. Mater. 105 (2016) 218–226.
[8] V.I.E. Ajiwe, C.A. Okeke, F.C. Akigwe, A preliminary study of manufacture of cement from rice husk ash, Bioresour. Technol. 73 (2000) 37–39.
[9] Specification for structural clay load-bearing tile.TIS:102-2517,1974, Bureau of Thai Industrial Standard, Bangkok.
[10] Specification for interlocking block (load-bearing type).TCPS:602-2547,2004, Bureau of Thai Industrial Standard, Bangkok.
[11] United States Environmental Protection Agency. Method 1311: toxicity characteristic leaching procedure 1992, SW-846, 1992

Acknowledgements
The author would like to express gratitude to researchers, namely Dr. Rattanakorn Rawangkul, STRR Engineering CO.,LTD, Thailand, Assoc. Prof. Dr. Wanchai Yodsudjai, Head of Department of Civil Engineering, Kasetsart University, Assoc. Prof. Dr.Trakool Aramraks, Department of Civil Engineering, Kasetsart University, Thailand for their invaluable help and constant encouragement throughout this research. Sincere thanks to the Department of Civil and Architecture for instrument test supporting and Muban Chombueng Rajabhat University, MCRU for partial financial support.

Measurement system analysis of viscometers used for drilling mud characterization

M S Mat-Shayuti, S N Adzhar

Faculty of Chemical Engineering, Universiti Teknologi MARA, Shah Alam, Malaysia

Email: mshafiq5779@salam.uitm.edu.my

Abstract. Viscometers in the Faculty of Chemical Engineering, University Teknologi MARA, are subject to heavy utilization from the members of the faculty. Due to doubts surrounding their result integrity and maintenance management, Measurement System Analysis was executed. 5 samples of drilling muds with varied barite content from 5 – 25 weight% were prepared and their rheological properties determined in 3 trials by 3 operators using the viscometers. Gage Linearity and Bias Study were performed using Minitab software and the result shows high biases in the range of 19.2% to 38.7%, with non-linear trend along the span of measurements. Gage Repeatability & Reproducibility (Nested) analysis later produces Percent Repeatability & Reproducibility more than 7.7% and Percent Tolerance above 30%. Lastly, good and marginal Distinct Categories output are seen among the results. Despite acceptable performance of the measurement system in Distinct Categories, the poor results in accuracy, linearity, and Percent Repeatability & Reproducibility render the gage generally not capable. Improvement to the measurement system is imminent.

1. Introduction

A bad gage will lead to wrong and perhaps costly decision [1]. Measurement system always increases the total observed variability of the parts measured. Total variability is equal to process variability plus measurement variability. While process variability only relates to variation caused by production process, raw material, equipment condition and environment, measurement variation on the other hand is a variation due to gage. Variation due to gage consists of four components which are accuracy, linearity, stability and precision.

MSA is a statistical tool specifically designed to determine the amount of variation contributed by measurement system [2]. It examines the collection of measurement tool, procedures, operations, software and operators that influence a measurement characteristic. MSA is capable of measuring gage characteristics such as accuracy, bias, and precision [3]. The importance of MSA in manufacturing industry is reiterated by many researchers [4-6]. Industry must use valid measurement data to ensure quality measurement results. As with the case of any measurement system, proper data collection and capability evaluation are needed to meet the measurement requirements in production and to improve quality control. MSA is currently practiced by many giants in manufacturing industry such as General Motor Corporations, Ford Motor Company, and Chrysler Group LLC to achieve required reliability of the production measurement [7].

Reports in MSA application and measurement system improvement are abundance. In 2010, Wu, Wei, and Li [8] evaluated the application of MSA in quality control process. The study laid down the findings of bias, linearity, stability, and precision in the experimental platform of large electronic current in manufacturing environment. In another research, Huang and Wang [9] studied the factors of measurement deviations across different operators. Operators were observed and recorded during the gage reading with causes affecting the judgment were examined. The causes then were mitigated with standardized operating procedures and the variation later was reduced, proving the effectiveness of MSA. Then there was a proposal made by Hajipoura et al. [10] to analyze measurement systems under fuzziness of indices in order to make the environment more realistic. Fuzzy measurement system analysis with gage repeatability and reproducibility (GR&R) index as a triangular fuzzy number was suggested. The proposed method was validated in a case study in automotive parts industry to display the impact of the innovation.

Apart from industrial applications, analysis of experimental equipment in learning institution is not uncommon as "the quality level of experimental devices impacts experimental teaching in colleges and universities directly" [11]. Therefore, this project was conducted to investigate the accuracy and precision of the viscometers in the Drilling and Geology Laboratory, Faculty of Chemical Engineering, Universiti Teknologi Mara (UiTM). The viscometers in focus are greatly utilized for drilling mud characterization by students and researchers. In drilling mud formulation, decision on process change depends a lot on measurement. From the interview with lab technician [12], the viscometers undergo scheduled maintenance once in three months. On top of that, he claimed records of usage, calibration, maintenance and any relevant certificates are all in place. However, this is highly doubted since there are no logbooks near the equipment to support that claim, thus, led to this investigation. Furthermore, lots of academic data have been coming out from the viscometers and published in scientific reports, making it even more vital to ensure that the equipment produce valid data. In this research, MSA methodology and analysis was accessed and generated from statistical software Minitab. Minitab was chosen as it finds wide applications in manufacturing industry as well as in education and research [13-14].

2. Methodology

2.1. Materials
The drilling fluid used was water-based mud with barite and bentonite as additives. Two units of viscometers under investigation were Grace Instrument viscometer model M3600 and Fann viscometer model 35SA 6-Speed. Fann viscometer was set at standard rotor R1, bob B1, and torsion spring F1 operating at 300 RPM.

2.2. Mud preparation
5 samples were prepared to cover the normal range of drilling mud, each with different percentage of barite. Barite was used as a weighting agent to increase the mud density [15]. Table 1 shows the composition for each sample. After mixing process, the muds then were characterized by the measurement system in place.

Table 1. Composition of drilling mud samples.

Mud samples	Percentage of barite (weight %)	Volume of water (mL)	Mass of bentonite (g)	Mass of barite (g)
S1	5	400	32.00	22.74
S2	10	400	32.00	48.00
S3	15	400	32.00	76.24
S4	20	400	32.00	108.00
S5	25	400	32.00	144.00

The first rheological property, plastic viscosity (PV), is a flow resistance of a fluid caused by mechanical friction within the fluids itself with solid particles [15]. The equation is given as

$$PV = \theta_{600\ RPM} - \theta_{300\ RPM} \quad (1)$$

where θ_N is the viscometer dial reading at any rotational speed of N. The intersection between shear stress and shear rate is known as yield point (YP) and is given by

$$YP = \theta_{300\ RPM} - PV \quad (2)$$

The final mud property is 10 min-gel strength (GS). GS can be obtained by taking the reading of drilling fluid at 600 RPM and 3 RPM. The drilling fluid was stirred at 600 RPM for 30 seconds. Then, the drilling fluid was left to rest for 10 minutes, before stirred again at 3RPM, and the very first peak reading was the GS.

On the other hand, characterization of the standard samples was outsourced to SCOMI Group Bhd., an international drilling service company with qualified laboratory. The data for the standard drilling mud samples is tabulated in the table 2.

Table 2. Standard sample values

Sample	Percentage of barite (wt%)	Plastic Viscosity, PV (cP)	Yield Point, YP(cP)	10 mins- Gel Strength, GS (lb/100ft^2)
S1	5	7	52	50
S2	10	9	54	61
S3	15	10	42	65
S4	20	11	56	73
S5	25	13	56	86

3. Measurement System Analysis (MSA)

3.1. Accuracy and linearity study

Gage Linearity and Bias Study analysis in Minitab was run using the collected data. Bias is a measure of accuracy of a set of measured values against the standard values of the characteristic being measured. To calculate bias

$$\text{Bias} = X - \mu \quad (3)$$

where X refers to the mean of the experimental data while μ is mean of the standard sample. After that, linearity was computed through the slope of mean error versus true value plot. Linearity is the difference in bias throughout the expected range of measurements, and is given by

$$\text{Linearity} = \text{Slope x Process Variation} \quad (4)$$

However, since Process Variation can only be obtained from historical data, Study Variation (or standard deviation multiplied by six) from the GR&R analysis was used.

3.2. Precision study

Gage Repeatability & Reproducibility (Nested) analysis in Minitab was chosen since the experiment was a destructive test in nature. The study was initiated by calculating MSA matrix to decide on the number of samples, trials and operators. The equation of MSA matrix is

$$(\#\text{samples}) \text{ x } (\#\text{operators}) \text{ x } (\#\text{trials} -1) \geq 30 \text{ (minimum)} \quad (5)$$

30 is the minimum number of required MSA matrix that is widely applied in the manufacturing industry [16]. To meet this requirement, 5 samples, 3 operators, and 3 numbers of trials were decided.

IOP Conf. Series: Materials Science and Engineering **222** (2017) 012003 doi:10.1088/1757-899X/222/1/012003

Operators were randomly selected among Oil and Gas Engineering students. No specific training was conducted and they were asked to complete the task by using their existing knowledge and experience. The experiment then took place according to the order generated by Minitab.

Precision study can be broken down into two parts, repeatability and reproducibility. From this study, Percent Repeatability and Reproducibility (%R&R), Percent Tolerance (%Tolerance) and Distinct Categories (DC) were determined. %R&R is given by %Tolerance was calculated by dividing the Study Variation for each component with process tolerance.

$$\%R\&R = \frac{\sigma^2}{\sigma^2} \; or \; \frac{\sigma^2_{Gage}}{\sigma^2_{Parts}} \tag{6}$$

The tolerance range for PV, YP, and GS were 5-15 cP, 39 – 61 cP, and 48 – 88lb/100ft^2, respectively. These ranges were the assumption of lower and upper tolerance for each drilling mud property, made from the normal operating range of the measurement system. Therefore

$$\%Tolerance = \frac{5.15\,\sigma_{Measurement}}{Tolerance} \; or \; \frac{5.15\,\sigma_{Gage}}{Tolerance} \tag{7}$$

where 5.15 is the standard deviations that accounts for 99% of measurement system variation and it is used as an industry standard [16].

Lastly, DC is a smallest detectable increment between the two measured values. It is given as

$$DC = \frac{\sigma^2_{Parts} * 1.41}{\sigma^2_{Gage}} \tag{8}$$

4. Result and Discussion

4.1. Gage linearity and bias study

Among the drilling mud properties, the lowest bias is shown in PV analysis with average bias percentage of 19.2%. GS follows closely at 24.7% and finally YP gives 38.7%. These data indicate incapable measurement system. In terms of linearity, the gage is poor as the highest linearity shown is only 54.2% for PV and thelowest is 8.3%. This suggests that the accuracy of the measurement system is not linear throughout the range of measurements, as some locations show better accuracies than others. The example of plot for Gage Linearity & Bias Study from Minitab is shown in figure 1.

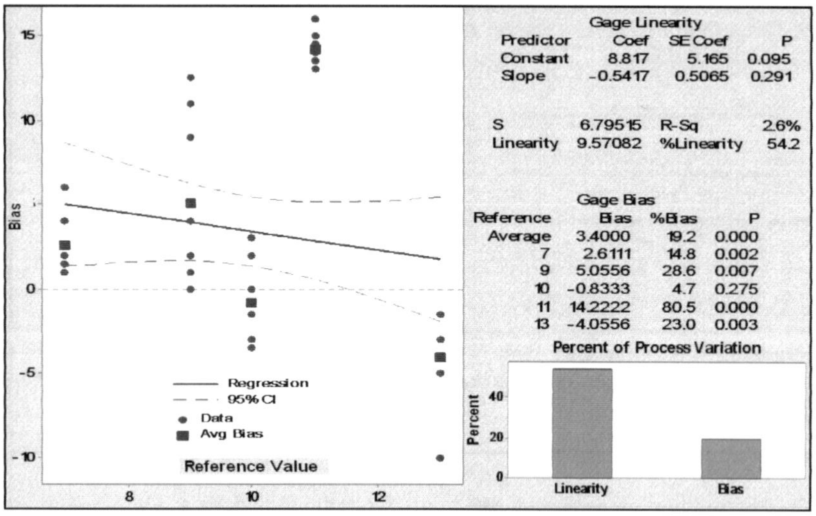

Figure 1. Gage Linearity and Bias for PV.

4.2. Gage repeatability & reproducibility (nested) analysis

From figure 2, in terms of total %R&R, YP portrays marginal performance at 5.11%. This is followed by poor indicator of GS with 8.05% and PV with 15.85%. All the %R&R contributions come from repeatability alone. The reproducibility contribution shows 0% contribution, indicating that operators' influence in gage variation is very small or none. This is similar to a finding by Špeťuch et al. [17] who studied the MSA of viscometers for vegetable and mineral oils. They found that the operators did not account for statistically significant outcome on the measured viscosity. Repeatability is the ability of a gage to repeat a measurement on the same sample under the same operating conditions while reproducibility is the ability to reproduce a measurement on the same unit under different operators. The standard range for good %R&R is from 0-2%, and over 7.7% is considered a bad range [16].

For %Tolerance, the lowest is displayed for GS at 54.82%, YP at 70.78%, and finally PV at 176.69%. %Tolerance is the ratio of precision to tolerance, with lower value is preferred as it suggests that the impact of variation due to gage precision is low with the given tolerance range. Since GS has the highest tolerance it is the likeliest to have lowest %Tolerance and vice versa. The industry-recommended value for &Tolerance is below 10% [16] and thus gage is deemed incapable for PV, YP and GS.

Finally, DC for PV, YP, and GS are 3, 6, and 4, respectively. This suggests that the gage is most favourable to YP, GS, and PV, in this order. It is good for YP, and marginal for GS and PV. Below DC of 3, the gage would be regarded as poor. DC means the number of groups within process data that measurement system can discern.

Generally, through %R&R and %Tolerance, it can be deduced that the measurement system in focus is not capable for drilling mud characterization. Improvement towards the gage needs to be carried out. The example of graphical representation of Gage Repeatability & Reproducibility (Nested) analysis in Minitab for PV is as shown in figure 3.

Gage R&R (Nested)	**Plastic Viscosity**					**Yield Point**					**Gel Strength**				
Source	DF	SS	MS	F	P	DF	SS	MS	F	P	DF	SS	MS	F	P
Operators	2	1.30	0.650	0.0044	0.996	2	22.40	11.217	0.0139	0.986	2	3.74	1.872	0.0040	0.996
Parts (Operators)	12	1761.83	146.819	16.9299	0.000	12	9690.30	807.52	56.6682	0.000	12	5651.07	470.922	35.2604	0.000
Repeatability	30	260.17	8.672			30	427.50	14.25			30	400.67	13.356		
Total	44	2023.30				44	10140.20				44	6055.48			

Gage R&R						
Source	VarComp	%Contribution (of VarComp)	VarComp	%Contribution (of VarComp)	VarComp	%Contribution (of VarComp)
Total Gage R&R	8.6722	15.85	14.2500	5.11	13.3560	8.05
Repeatability	8.6722	15.85	14.2500	5.11	13.3560	8.05
Reproducibility	0.0000	0.00	0.0000	0.00	0.0000	0.00
Part-To-Part	46.0491	84.15	264.4240	94.89	152.5220	91.95
Total Variation	54.7213	100.00	278.6740	100.00	165.8780	100.00
Process Tolerance	10		32		40	

Souce	StdDev (SD)	Study Var (6*SD)	%Study Var (%SV)	%Tolerance (SV/Toler)	StdDev (SD)	Study Var (6*SD)	%Study Var (%SV)	%Tolerance (SV/Toler)	StdDev (SD)	Study Var (6*SD)	%Study Var (%SV)	%Tolerance (SV/Toler)
Total Gage R&R	2.94486	17.6692	39.81	176.69	3.77490	22.6500	22.61	70.78	3.65450	21.9272	28.38	54.82
Repeatability	2.94486	17.6692	39.81	176.69	3.77490	22.6500	22.61	70.78	3.65450	21.9272	28.38	54.82
Reproducibility	0.00000	0.0000	0.00	0.00	0.00000	0.0000	0.00	0.00	0.00000	0.0000	0.00	0.00
Part-To-Part	6.78595	40.7157	91.73	407.16	16.26110	97.5670	97.41	304.90	12.35000	74.0999	95.89	185.25
Total Variation	7.39738	44.3843	100.00	443.84	16.69350	100.0000	100.00	313.00	12.87940	77.2761	100.00	193.19
Distinct Categories =	3				6				4			

Figure 2. Gage R&R (Nested) report for PV, YP, and GS.

4.3. Cause analysis for measurement system variation

There are a few possible factors contributing to this overall poor measurement system capability. Firstly, the maintenance program for the viscometers is suspicious. There is no strong proof of maintenance plan and execution in place. Without it, wear and tear, calibration, and even safety of the gage are questionable.

Secondly, there is no standard operating procedure (SOP) established for the viscometers. It is important to ensure proper handling when dealing with viscometers' speed setting and choosing the right rotor. SOP also helps in judgement when reading analog display, recognizing peak in dial reading, and providing correct formulas for PV, YP, and GS calculations.

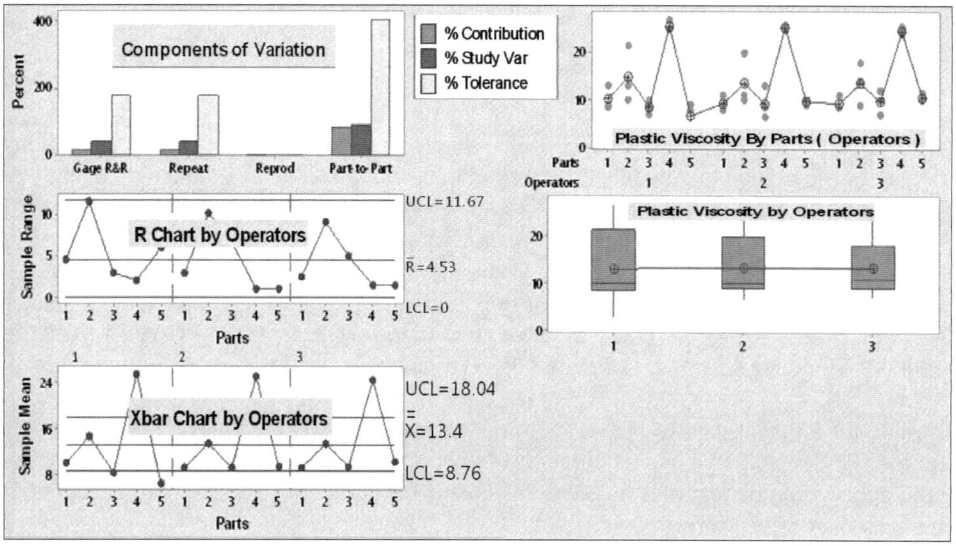

Figure 3. Gage R&R (Nested) plot for PV.

5. Conclusion

This work is about MSA for drilling mud characterization in the Faculty of Chemical Engineering, UiTM. This measurement system mainly involves the viscometers and operators within a setting of the Drilling & Geology Laboratory. Through the Gage Linearity & Bias study, biases values range from 24.7% to 38.7%. The bias also portrays non-linear trends throughout the range of measurements. From the Gage Repeatability & Reproducibility (Nested) study, %R&R is marginal for YP but poor for GS and PV. The gage is then estimated to be incompetent for all mud properties as they exhibit more than 30% in %Tolerance. Lastly, DC is good for YP but marginal for GS and PV determination. Overall, the findings definitely call for the measurement system to be improved.

6. Recommendation

It is a good practice to start logging measurement data in order to measure gage stability through control chart. From it, causative factors for variability may be discovered and the frequency of calibration can be planned accordingly to minimize error due to instability. Besides, MSA will become more accurate with historical Process Variation available. Furthermore, it is suggested that blind or random sampling is applied to eliminate noise in analysis

Improvement to the measurement systems is definitely required until accuracy, %R&R, %Tolerance, and DC fall within good ranges. Apart from keeping a logbook and sound maintenance/calibration plan, SOP for the viscometers will be helpful to reduce error due to measurement system.

References

[1] Kazerouni A 2009 *World Acad of Sc., Eng. & Tech.* **52** 31-35
[2] Lewis P, Cooke G 2013 *Int. J. Metrol. Qual. Eng.* **4** 145-51
[3] Pai F, Yeh T, Hung Y 2015 *Sustainability* **7** 15464-86
[4] Wu H 2013 *Appl. Mech. & Mat.* **455** 527-32
[5] Senol S 2004 *Measurement* **36** 31–141
[6] Larson G 2003 *Quality Eng* **16** 297-306
[7] Down M, Czubak F, Gruska G, Stahley S, Benham D 2010 *Measurement Systems Analysis*

Reference Manual 4th Edition (Michigan: MSA Work Group)

[8] Wu H, Wei X, Li W 2010 *Key Eng. Mat.* **426-427** 55-59

[9] Karwowski W, Salvendy G 2010 *Advances in Human Factors, Ergonomics, and Safety in Manufacturing and Service Industries* (Florida: CRC Press)

[10] Hajipoura V, Kazemib A, Mousavib S 2013 *Measurement* **46** 2770–80

[11] Hong C, 2011 Application of MSA in quality evaluation of experimental devices *Res. & Expl. in Lab.* ***1***

[12] Yusof M, personal communication, May 2016.

[13] Al-Refaie A, Bata N 2010 *Measurement* **43** 842-51

[14] Brown R 1985 *Beh. Res. Meth., Instr., & Comp.* **17**, 573–75

[15] Dhiman A 2012 *Rheological Properties & Corrosion Characteristics of Drilling Mud Additives* (Nova Scotia: Dalhousie University)

[16] Kappele W, Raffaldi J 2010 *Quality Magazine* **5** 32-34

[17] Špeťuch V, Petrík J, Grambálová E, Medveď D, Palfy P 2015 *Acta Metallurgica Slov.* **21** 53-60

Acknowledgement

This research was under the funding of Research Acculturation Grant Scheme (RAGS/1/2014/TK04/UITM//8) by Ministry of Education Malaysia through Faculty of Chemical Engineering, Universiti Teknologi MARA, Selangor.

IOP Conf. Series: Materials Science and Engineering **222** (2017) 012004 doi:10.1088/1757-899X/222/1/012004

Thermal diffusivity and adiabatic limit temperature characterization of consolidate granular expanded perlite using the flash method

Saad Raefat[1,b], Mohammed Garoum[1,a,*], Najma Laaroussi[1], Macodou Thiam[2] and Khaoula Amarray[1].

[1] Materials, Energy and Acoustics TEAM, Higher School of Technology Salé, Mohammed V University in Rabat, 11060, 227 Avenue Prince Héritier, Morocco.
[2]University Institute of Technology of the University of Thiès, Senegal.

[a]garoum1@yahoo.fr,[b]saad.raefat@gmail.com

Abstract: In this work experimental investigation of apparent thermal diffusivity and adiabatic limit temperature of expanded granular perlite mixes has been made using the flash technic. Perlite granulates were sieved to produce essentially three characteristic grain sizes. The consolidated samples were manufactured by mixing controlled proportions of the plaster and water. The effect of the particle size on the diffusivity was examined. The inverse estimation of the diffusivity and the adiabatic limit temperature at the rear face as well as the heat losses coefficients were performed using several numerical global minimization procedures. The function to be minimized is the quadratic distance between the experimental temperature rise at the rear face and the analytical model derived from the one dimension heat conduction. It is shown that, for all granulometry tested, the estimated parameters lead to a good agreement between the mathematical model and experimental data.

1. Introduction

In order to guarantee a better thermal performance inside buildings decrease their high energy demand and ensure a better occupant's comfort, the Moroccan Thermal Regulation of Construction (MTRC) has become mandatory in November 2015. Indeed, its establishment requires creating favorable conditions for its implementation. For this purpose, this work aims to improve the thermophysical properties of one Moroccan building material. Accordingly, several samples were prepared mixing the gypsum plaster with granular expanded perlite; some of their thermophyisical properties were then investigated on three different granular size. The thermal diffusivity was determined using the flash

method and the adiabatic temperature at the rear face of the sample was estimated using several numerical global minimization procedures.
Recent studies on mixing expanded perlite with building materials were investigated such as in [1] studied the effect of expanded perlite on the mechanical properties and thermal conductivity of lightweight concrete or in [2] who compared the thermal properties of clay when it's embedded with expanded perlite or granular cork. Reference [3] also determined an experimental and numerical investigation of a hollow brick filled with perlite. Moreover [4] addressed the thermal properties of syntactic metal foam made by embedding expanded perlite particles in aluminum.

The present study aims to estimate the thermophysical properties of expanded perlite mixed with gypsum plaster.

2. Description of used materials

2.1 Expanded perlite.

The material subject of work is the granular expanded perlite (Figure 1); it was taken from Berchid's industrial zone located in the northwestern part of Morocco. The perlite is a siliceous volcanic natural rock, family of the pearlitic rhyolites. At first, the rock is crushed, and calibrated by particle sizes. Industrial expansion perlite is carried out in special ovens, stationary or rotating. Under the effect of heat (up to 1200 °C), the perlite grains expand: a multitude of closed cells are formed inside the grains. Indeed, the steam expands the material to 15 times its original volume. For this reason, it is mainly used in new construction and rehabilitation as lightweight concrete and aggregate for floors of any kind. In panel form, it is suitable as a sealing support to all types of roofs and all environmental medium.

2.2 Gypsum plaster

The gypsum plaster used in this comparison study is manufactured in Morocco, more precisely, in Safi's industrial zone. It is used to protect the external walls from moisture and damp, the internal walls from the influence of water vapor, and it plays an esthetic role as well.

3. Experimental approach

3.1 Samples preparation

Three samples were prepared corresponding to three granular expanded perlite categories: C1 [0.625mm; 2.5mm], C2 [2.5mm;5mm] and C3 [5mm; 8mm] shown in Figure1. Particle size distribution of the expanded perlite is presented in Table1. In order to compare the effect of expanded perlite on the thermophysical properties of the samples, a fourth one was prepared using only gypsum plaster. All the samples were formed in a circular mold with 10.3cm in diameter d and 2cm in thickness e. The full mold apparent volume was filled with expanded perlite, right after a certain volume of gypsum plaster was added until that the inter-granular space is entirely occupied. The four samples were covered by a thin black painting, and then introduced into a stove for two days. To ensure that all the pores moisture is totally removed, they were weighted several separate times until their mass became constant. Finally they were packed in plastic bags to maintain zero humidity until experimental measurements use. Water to gypsum plaster ratio was kept equal to 100% in all mixtures. The four samples are presented in Figure 2.

Table 1: Aggregate grading of the used expanded perlite

Sieve size [mm]	8	5	2.5	0.63
Percentage passing	100%	97%	57%	4%

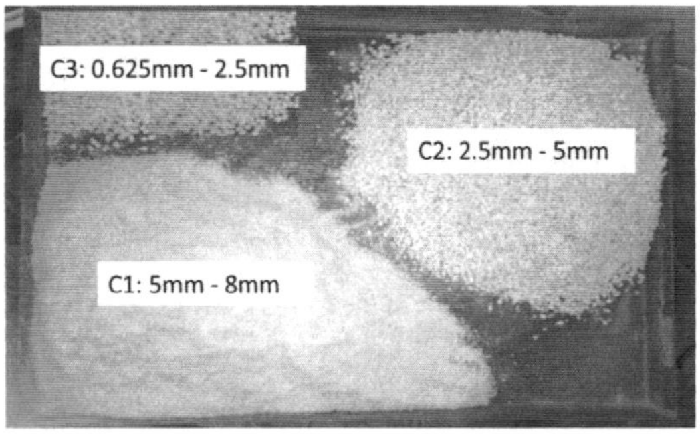

Figure 1. The three loose granular perlite from which samples were prepared.

Figure 2. The four samples tested PP0: plaster alone; PP1: mix using C1, PP2: Mix using C2; PP3: mix using C3

3.2 The flash method

The flash method is adopted for thermal diffusivity determination of many materials. It's based on the emission of an energy pulse parallel to the tested material front face; and for a very short time, a thermocouple in contact with the rear face enables the record of the temperature increasing in function of time once the front face has received the flash. Its schematic is shown in Figure 3.

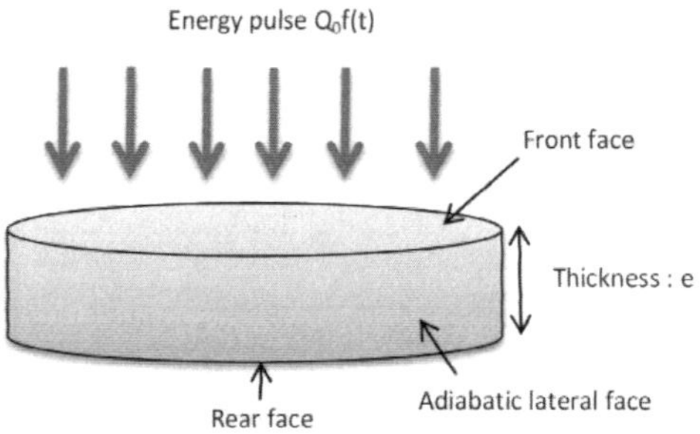

Figure 3. Schematic of the flash method.

Once the steady state is reached the flash is started manually and the temperature of the rear face is recorded with adequate sampling rate. The same procedure is repeated many times on the same sample until three satisfactory curves are obtained.

4. Theoretical approach and global numerical minimisation

A modeling has been done to estimate the thermal diffusivity and the adiabatic limit temperature using the global minimization. Assuming the uniformity of the absorbed energy and an initial temperature T_0 equal to the ambient temperature, the rise temperature T is governed by the following system.

$$\begin{cases} \dfrac{\partial^2 T(x,t)}{\partial x^2} = \dfrac{1}{a}\dfrac{\partial T}{\partial t} = 0. \\ \dfrac{\partial T(x,t)}{\partial x}\Big|_{x=0} = -h_1.T(0,t) + Q_0.f(t). \\ \dfrac{\partial T(x,t)}{\partial x}\Big|_{x=e} = h_2.T(e,t). \\ f(t) = \begin{cases} \dfrac{1}{t_d} & , \; 0 \le t \le t_d. \\ 0 & , \; t_d \le t. \end{cases} \\ T(x,0) = 0. \end{cases} \tag{1}$$

a is the thermal diffusivity, h_1 and h_2 are the global heat exchange coefficients on both sides of the sample, $Q_0.f(t)$ is the finite pulse with duration t_d, Q_0 (W/m^2) is the finite amount of heat absorbed at the front boundary (x=0) and f(t) is the time dependence of the heat generation. For all measurements t_d is kept equal to 2s. This duration is sufficient to reach more than 1 °C of the maximum rise temperature at the rear face.

The Laplace transform of Eq. 1 allows expressing the solution of the temperature in the Laplace space as:

$$T(e,p) = T_{ma}.F(p)\frac{e^2}{a}.\frac{1}{\left[\dfrac{(b_{i1}.b_{i2})A(p)}{\sqrt{e^2/a}} + (b_{i1}+b_{i2})B(p) + A(p)\sqrt{e^2/a}\right]}. \tag{2}$$

With:

$$A(p) = \text{Sinh}\left(\sqrt{\frac{e^2}{a}p}\right). \qquad B(p) = \text{Cosh}\left(\sqrt{\frac{e^2}{a}p}\right). \tag{3}$$

e is the thickness of the sample. b_{i1} and b_{i2} are the Biot numbers respectively.

$$b_{i1} = \frac{h_1 e}{\lambda}. \qquad b_{i2} = \frac{h_2 e}{\lambda}. \tag{4}$$

F(p) is the Laplace transform of $f(t)$ expressed as:

$$F(p) = \frac{1 - e^{-p.t_d}}{p.t_d}. \tag{5}$$

$T_{ma} = \dfrac{Q_0}{\rho ce}$ is the adiabatic limit temperature. Where ρ is the density and c is the specific heat.

The time dependence of the temperature on the rear face $T_{th}(e,,T_{ma},a,b_{i1},b_{i2},t)$ is then obtained by inversing numerically the solution in Eq. 2. Several algorithms were tested and compared (Papouilis, Talbot, Piessen, Crump, DeHoog, Gaver–Stehfest) ; among them the numerical method Gaver–Stehfest[5] was chosen for its swiftness and its easier numerical implementation. A special script was written in Mathematica language.

In order to characterize the samples thermally, parameters related to the theoretical model predicting T_{th} namely T_{ma}, the diffusivity a and the Biot numbers b_{i1} and b_{i2}were simultaneously estimated from experimentaldata. The optimal values of these parameters were simultaneously estimated by minimizing the quadratic distance M between the theoretical model stemming from the numerical inversion and the experimental thermogram.

$$M(e,T_{ma},a,b_{i1},b_{i2}) = \sum_{(i=1)}^{N}\left[T_{exp}(t_i) - T_{th}(e,T_{ma},a,b_{i1},b_{i2},t_i)\right]^2. \tag{6}$$

t_i is the time points and N is the length of the experimental data vector taken into account.

To solve the minimization problemin Eq. 6 of the multimodal function M, the Nelder Mead algorithm was implemented using Mathematica language.

5. Results and discussion

5.1 Density:

Knowing the samples exact dimensions and dry masses, the apparent density can be easily deduced, all these parameters are presented in Table 2.

Table 2: Samples dimensions and density

Sample	e [cm]	d [cm]	m [g]	ρ [kg/m^3]
PP0	2.00	10,2	140,9	215.54
PP1	1.95	10,1	112,5	176.51
PP2	1.95	10,3	102,7	158.02
PP3	2.00	10,2	090,4	138.28

pp1, pp2 and pp2 represent the plaster mixed with perlite for the three granular expanded perlite categories: [0.625mm; 2.5mm], [2.5mm; 5mm] and [5mm; 8mm] respectively.

It is clear from Table 2 that the higher the expanded granular perlite size is the lower is the apparent density of the sample. The density decreased from 216 kg/m^3 for plaster alone up to 138 kg/m^3 for the biggest perlite category C3 (PP3 sample), with 36% of gain on lightness.

5.2 Thermal diffusivity and adiabatic limit temperature:

Table 3: Predicted thermal diffusivity and adibatic limit temperature

Sample	Test	Estimated diffusivitya 10^{-7} [m^2.s^{-1}]			Estimated T_{ma} [°C]
		Degiovanni [6]	Parker [7]	Present study	Presentstudy
PP0	1	2.92	3.16	2.79	1.66
	2	2.99	3.24	2.90	1.57
	3	2.92	3.16	2.79	1.66
	Mean value	2.94	3.19	2.83	1.63
PP1	1	2.79	3.16	2.80	2.00
	2	2.80	3.17	2.80	2.00
	3	2.93	3.20	2.92	2.00
	Mean value	2.84	3.18	2.84	2.00
PP2	1	2.84	3.24	2.96	2.00
	2	2.86	3.27	2.91	2.00
	3	2.98	3.44	3.11	2.00
	Mean value	2.89	3.32	2.99	2.00
PP3	1	3.23	3.73	3.53	2.00
	2	3.12	3.70	3.55	2.00
	3	3.14	3.70	3.67	2.00
	Mean value	3.16	3.71	3.58	2.00

Table 3 shows the thermal diffusivity of the four samples using three models; the Parker and Degiovanni known models and the proposed global minimization, and the adiabatic limit temperature using the numerical inversion.

From Table 3, the thermal diffusivity values estimated from the Global minimization method are close to those of obtained using the Degiovanni procedure; moreover, these values are included between the Degiovanni and Parker estimated values.

From the analysis of the adiabatic limit temperature, it is clear that it increases for mixtures compared to the plaster alone, which means that ρc decreases. However, for all different mixtures studied T_{ma} remains constant, which means that ρc also remains constant. We can deduce that c is increasing with the increase of the particle size since the density decreases (Table 2).

The estimated parameters are then injected into T_{th} in order to compare theoretical and experimental data. Figure 4 presents the predicted and measured thermogram for the four samples. It is interesting to note the well agreement between T_{th} with predicted values of parameters and experimental data.

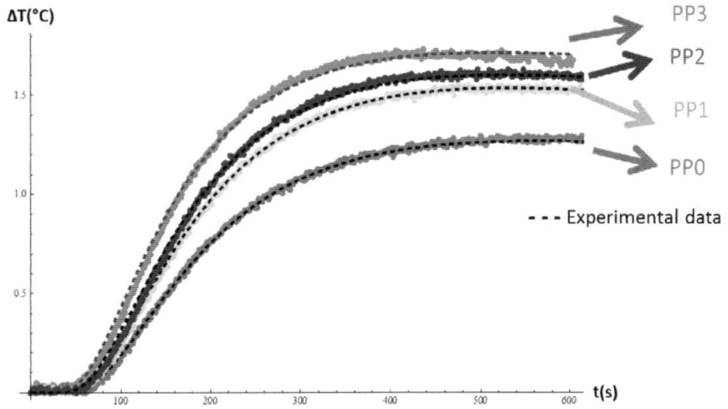

Figure. 4. Comparison between the theoretical model and experimental data.

6. Conclusion

This paper has presented an experimental study of the new material gypsum plaster embedded with expanded granular perlite using the flash method. Three thermophysical properties were returned, the density was directly deduced knowing the samples dimensions and masses, the thermal diffusivity was determined through three different mathematical models and the adiabatic limit temperature was estimated using the global minimization procedure. It was concluded that with the increase of the perlite particle size, the thermal diffusivity along with the thermal capacity increased, and the density decreased, this might be due to the water absorption of the expanded perlite particles. Moreover, a good agreement between T_{th} with predicted values of parameters and experimental data was obtained. As perspective, a second numerical procedure is in progress in order to examine precisely the thermal conductivity and the specific heat using only the flash method.

References

[1] O. Sengul, S. Azizi, F. Karaosmanoglu, and M. A. Tasdemir, "Effect of expanded perlite on the mechanical properties and thermal conductivity of lightweight concrete," *Energy Build.*, vol. 43, no. 2–3, pp. 671–676, 2011.

[2] Y. Maaloufa, S. Mounir, A. Khabbazi, J. Kettar, and A. Khaldoun, "Thermal Characterization of Materials based on Clay and Granular: Cork or Expanded Perlite," *Energy Procedia*, vol. 74, pp. 1150–1161, 2015.

[3] M. Zukowski and G. Haese, "Experimental and numerical investigation of a hollow brick filled with perlite insulation," *Energy Build.*, vol. 42, no. 9, pp. 1402–1408, 2010.

[4] T. Fiedler, I. V. Belova, and G. E. Murch, "On the thermal properties of expanded perlite – Metallic syntactic foam," *Int. J. Heat Mass Transf.*, vol. 90, pp. 1009–1014, 2015.

[5] Stehfest, Numerical inversion of Laplace transforms. Communications of the Association for Computing Machinery, 1970.

[6] Degiovanni A., Laurent A., « Une nouvelle technique d'identification de la diffusivité thermique pour la méthode flash », Revue de Physique Appliquée, n° 21, pp. 229-237, 1986.

[7] Parker W.J., Jenkins R.J., Butler C.P., Abbot G.L., Flash method of determining thermal diffusivity, heat capacity and thermal conductivity , Journal of Applied Physics, vol. 32, n°9, pp. 1679-1684, 1961.

IOP Conf. Series: Materials Science and Engineering **222** (2017) 012005 doi:10.1088/1757-899X/222/1/012005

Effect of phase and size characteristics of fly ash from power station on properties of bitumen composites

M S Lebedev[1], N I Kozhukhova[1] and I L Chulkova[2]

[1] High Technology Center, Belgorod State Technological University named after V.G. Shoukhov, Kostukov St, 46, Belgorod, Russia
[2] Institution of Civil Engineering, Siberian State Automobile and Highway Academy, Mira Avenue, 5, Omsk, Russia

Email: michaelL1987@yandex.ru

Abstract. Fly ash from power stations is one of the effective fillers in bitumen binders due to its high specific surface area in origin state. This paper is focused on study of effect of phase and size characteristics of fly ashes on properties of bitumen binder. In this work a high-calcium (FA1) and low-calcium (FA2) fly ashes are used. According to chemical and mineral composition of FA1 is reactive in water medium because it contains hydraulic minerals. FA2 is hydraulically inert mainly consisting of aluminosilicate vitreous phase. FA1 is characterized by higher specific surface vs. FA2 according to Blain method data but lower specific surface measured with BET method. It can be associated with high porosity of FA2 that significantly influent on bitumen content in binding system. Lower packing density of FA2 particles leads to reducing of strength characteristics in bitumen composite. Low adsorption of bitumen to surface of FA1 particles initiates hydration process of reactive minerals as well as formation of ettringite phase that leads to extension of the binder system and structure decompaction. Bitumen composites based on FA2 with higher adsorptive capacity is characterized by more stable structure.

1. Introduction

At present time the interest to energy- and resource-saving technologies in construction industry including road area grows all over the word. To reduce the consumption of expensive bitumen binder as well as enhancement of thermal, mechanical characteristics and fatigue effect the introduction of high-disperse mineral powders and fillers into bitumen cement is applied.

The most popular mineral powders are from limestone and dolomite rocks. By-products from energy, metallurgical and chemical industry also are used, but their application is limited due to insufficient knowledge about effect in binder system taking into account variation in phase, size and structure properties. So, additional study in this area is required.

In this paper the high-disperse wastes from power stations – fly ashes – as the most suitable mineral fillers that intensively applied in bitumen binder [1]–[3] systems are studied. But, taking into account the insufficient understanding of fly ash features its effect in bitumen binders should be studied complexly to development the general concept of structure formation process and design the recommendation of its application.

Thus, in this work the influence of phase and size characteristics of fly ash on properties of bitumen composite is studied.

2. Experimental Part

2.1. Materials

Two types of fly ashes from Russian power station are used: high-calcium fly ash Nazarovsk power station (Krasnoyarsk Territory) –obtained by dry removing (FA1); low-calcium fly ash (particle size is 60 mm) from Omsk power station (Omsk) obtained by wet removing (FA2).

The construction viscous bitumen BND 60/90, (Russia) is used as organic binder (penetration is 60 dmm).

2.2. Methods

Chemical and mineral compositions of fly ashes are realized with XRF and XRF analysis, respectively measured by 9900 WorkStation with embedded diffraction system Thermo Fisher Scientific. Qualitative XRD-analysis is accomplished with database PDF–2 (ICCD) and by Full profile method with DDM v.1.95e program [4].

Dispersity of fly ashes is determined by following parameters:

– Size distribution measured with laser particle analyzer FRITSCH Analysette 22 NanoTec plus;
– Specific surface area measured by Blain method;
– Specific surface area measured by BET method with SORBI-M equipment.

Microstructure of fly ashes is studied by SEM analysis with electron microscope TESCAN MIRA 3 LMU.

3. Results and Discussions

3.1. Chemical and Mineral Compositions of Fly Ashes

The studied fly ashes are characterized by different chemical (Table 1) and mineral (Table 2) compositions due to different type of coal recourse and location of coal mining.

Table 1. Chemical composition of fly ashes.

Type of fly ash	Oxide content, wt.%									
	SiO_2	Al_2O_3	**CaO**	Fe_2O_3	MgO	SO_3	Na_2O	K_2O	TiO_2	LOI
FA1	31,55	8,84	**37,80**	8,99	6,31	4,40	0,76	0,20	0,26	3,15
FA2	55,33	29,92	**2,57**	7,48	1,90	–	0,45	0,42	1,10	8,29

According to data from Table 1 the lime factor M_0 is calculated by following:

$$M_0 = \frac{(CaO+MgO)}{(SiO_2+Al_2O_3)} \quad (1)$$

For FA1 parameter M_0=1.45 and for FA2 M_0=0.05. On the base of these data and in compliance with [5], [6] FA1 is high-calcium fly ash characterized by good reactivity in water medium and can be used as binding component. FA2 is low-calcium fly ash that inert in water and reactive in alkaline medium only (for example, geopolymer) and can be used as reactive mineral additive in cement. Mineral compositions of the studied fly ashes are differed also (Table 2).

Table 2. Mineral compositions of the fly ashes.

Mineral phase	Content, wt. %		Mineral phase	Content, wt. %		Mineral phase	Content, wt. %	
	FA1	FA2		FA1	FA2		FA1	FA2
Vitreous phase	49,26	49,97	Quartz	7,61	19,34	Mullite	–	26,36
Magnetite	1,56	3,49	Anhydrite	7,02	–	Calcite	0,69	–
Portlandite	3,47	–	CaO_{free}	4,09	–	C_3A	21,34	–
C_4AF	4,11	–	Periclase	0,85	–			

The FA2 consists of following crystal phases: quartz, mullite and magnetite i.e. mineral composition is typical for low-calcium fly ash [7]. The FA1 consists of following crystal phases including Ca-minerals: anhydrite ($CaSO_4$), CaO_{free}, portlandite ($Ca(OH)_2$), calcite ($CaCO_3$), tricalcium aluminate ($3CaO \cdot Al_2O_3$), tetracalcium alumoferrite ($4CaO \cdot Al_2O_3 \cdot Fe_2O_3$) and some of periclase (MgO).

All phases but calcite react with water resulting a new phase formation that explains a high value of M_o parameter (M_o=1,45).

According to quantitative XRD-analysis the vitreous phase content in both fly ashes is equivalent (Table 2). SEM-analysis (Figure 1) demonstrates the microsphere prevalence (especially in FA1) (Figure 1a) and presence of unburned carbon (Figure 1b). Also FA2 consists of anisotropic particles including porous inclusions and carbon residue. The values of LOI in the both fly ashes are significantly differed. That is associated with application of different coal recourses. Brown coal combustion initiates on formation of carbon residue up to 4 % (FA1), and bituminous coal combustion forms 3–12 % of carbon residue (in FA2) [6].

a b

Figure 1. Microstructure of fly ash particles: a – FA1; b – FA2.

So, vitreous phase of FA1 generally consists of microspheres. In case of FA2 the vitrous phase components are microspheres, carbon residue and anisotropic glass components.

3.2. Fly Ash Dispersity

Laser grain-size analysis demonstrates different dispersity of the fly ashes (Figure 2). For FA2 the size distribution is bimodal in range of 0.1–100 µm with dominant particle size of 20 and 45 µm. For FA1 the size distribution is unimodal in range of 0.1–50 µm with dominant particle size of 15 µm. For FA1 the more narrow range of particle distribution vs. FA2 is typical.

The higher dispersity of FA1 is confirmed by specific surface area data obtained with laser diffraction analysis and Blain method (Table 3). However specific surface area data determined with BET method is significantly higher for FA2 vs. FA1. It can be associated with microporosity of unborned carbon aggregates in FA2 (Figure 1) and leads to filler porosity in general as well as enhancement of adsorption capacity [6].

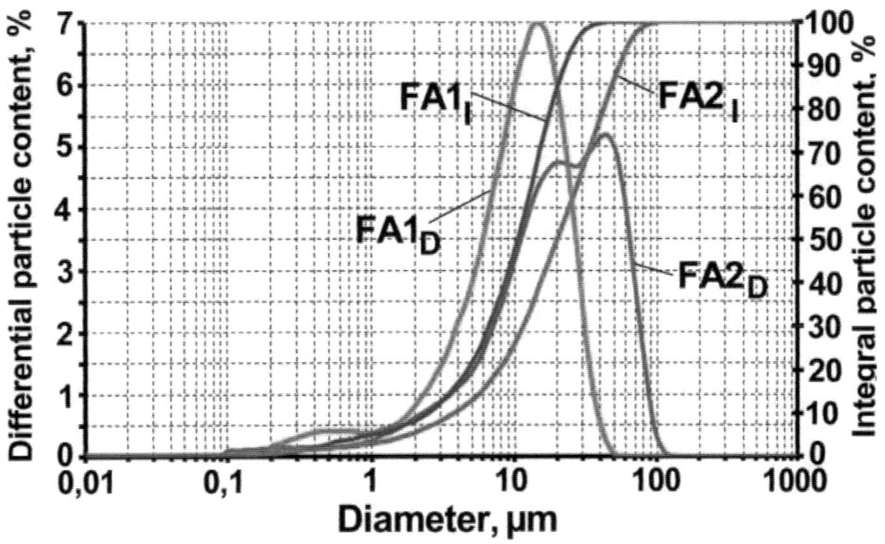

Figure 2. Differential (D) and Integral (I) curves of particle size distribution for: FA1– $FA1_D$ and $FA1_I$; FA2 – $FA2_D$ and $FA2_I$.

Table 3. Results of specific surface area for FA1 and FA2 obtained with different methods.

Type of fly ash	Specific surface area		
	by Blain method, m^2/kg	by BET method, m^2/kg	laser diffraction analysis m^2/kg (cm^2/cm^3)
FA1	347,4	1100±100	565,7 (15664)
FA2	291,6	5900±400	563,4 (11879)

3.3. Properties of Fly Ashes as Fillers in Bitumen Binder

Applied FA1 and FA2 are studied as mineral fillers MP-2 (non-carbonate rocks and industrial wastes) according to Russian Standard 52129–2003 [8] (Table 4).

Table 4. Basic characteristics of fly ash as mineral component in bitumen binders.

Parameter	According to Russian Standard 52129-2003 for mineral fillers MP-2	FA1	FA2
Real density, g/cm^3	–	2,78	2,10
Bulk density, g/cm^3 (under load of 40 MPa)	–	1,92	1,27
Porosity, % (under load of 40 MPa)	≤ 40	30,9	39,5
Humidity, wt. %	≤ 2,5	0,31	0,35
Bitumen content, g per 100 cm^3 of filler	≤ 80	47	74
Expansion of bitumen samples, vol. %	≤ 3,0	3,4	1,75
Water resistance of bitumen samples	<0,7	0,81	0,72

Porosity of fly ash depends on range of particle distribution, particle shape and microporosity. Also porosity of mineral filler influents on volume of free space in loaded material that hereafter will be filled by binding component when bitumen binder formation.

So, lower porosity value for FA1 vs. FA2 is associated with difference of bitumen content value (Table 4). The bitumen content parameter is "bitumen – filler" ratio provided the required viscosity of binding system. Difference in bitumen content values is higher (74 and 47 in FA1 and FA2, respectively) vs. difference in porosity values (41.4% and 36.3% under load of 10MPa in FA1 and FA2, respectively) that is connected with difference in specific gravity values for the fly ashes (Table

IOP Conf. Series: Materials Science and Engineering **222** (2017) 012005 doi:10.1088/1757-899X/222/1/012005

5). Relation between these parameters (bitumen content, porosity) is confirmed by earlier studies [9], [10].

According to Russian Standard 52129–2003 [8] the optimal bitumen content, providing the water saturation of bitumen binder samples of 4–5 % (cylinders with diameter and height of 25 mm pressed under load of 10 MPa) (Table 5).

Table 5. Properties of bitumen binders depending on component composition.

Parameter		FA1			FA2			
Bitumen content (of fly ash, wt. %)		18	**18,5**	19	18	24	36	**37**
Volume content of component, (unit fraction)	Fly ash	0,622	**0,623**	0,623	0,521	0,526	0,530	**0,531**
	Bitumen	0,308	**0,317**	0,326	0,195	0,262	0,397	**0,409**
	Pores	0,070	**0,060**	0,051	0,284	0,212	0,073	**0,060**
Water saturation, vol. %		6,15	**4,9**	3,5	28,1	21,0	5,9	**4,4**
Mass specific gravity of bitumen binder, g/cm^3 (under a load of 10 MPa)		2,040	**2,057**	2,065	1,291	1,368	1,514	**1,528**
Fly ash porosity in bitumen binder structure, % (under a load of 10 MPa)		37,8	**37,7**	37,7	47,9	47,4	47,0	**46,9**

According to data from Table 5 the optimal bitumen content (by wt. %) in FA2-bised bitumen binder is two times higher vs. FA1-bised bitumen binder (37 % vs. 18.5 %, respectively) providing the porosity values of 6 % in both cases. At the same time "fly ash – bitumen" ratio in the both binders is significantly different (1.96 and 1.3 for binders with FA1 and FA2, respectively) due to higher porosity of FA2 (46.9%) vs. FA1 (37.7%). Growing of bitumen concentration in the binder leads to the binder structure compaction until direct particle contact [11] and porosity reduction (Table 5).

Higher bitumen requirement in FA2-based binder is connected with higher surface area FA2 (according to BET method) and higher adsorption capacity vs. FA1 (Table 3).

Lower adsorption capacity of FA1 leads to layering of bitumen binder and hydration of Ca-mineral in the fly ash to ettringite formation that initiates expansion of bitumen binder samples of 3.4 % at least.

3.4. Strength Properties of Bitumen Binders with FA1 and FA2.

For optimal composition of bitumen binders the parameters of expansion, shirt-term water resistance and strength at different temperatures (0°C, 20°C, 50°C) are determined according to Russian Standard 52129–2003 [8]. Before the strength test the bitumen samples are thermostated for a hour. According to experimental data in Table 6 strength value of FA1-based bitumen binder is higher vs. FA1-based bitumen binder in all temperature range due to higher packing density of solids in the FA1-based bitumen binder according to characteristics of fly ash as well as bitumen binder porosity presented in Table 4 and 5, respectively. Data variations are in range of 16–20%.

Table 6. Strength characteristics of bitumen binders.

Parameter		FA1	FA2
Compressive strength, MPa	at 20°C	4,02	3,20
	at 0°C	6,77	5,68
	at 50°C	0,89	0,74
Heat-stability coefficient R_{20}/R_{50}		4,50	4,30
Thermal stability coefficient R_0/R_{50}		7,57	7,61

According to experimental data the heat-stability coefficient and thermal stability coefficient are calculated (Table 6). The resulting data are equivalent and demonstrate the possibility of application of optimal composition of fly ash based bitumen binder in equal temperature range.

4. Summary

The study results for FA1 and FA2 as mineral fillers in road bitumen composites demonstrate significant differences in structure formation, physical and mechanical properties of based construction materials. FA2 is characterized by high porosity and low particle packing density leading to growth of bitumen requirement and reducing of strength parameters vs. FA1.

At the same time FA1 has a better bitumen adsorption capacity providing more stable composite structure. Low adsorption capacity of FA1initiates layering process and structure decompaction under water effect.

So, effect of phase and size characteristics of mineral fillers including fly ash on bitumen composites is basic. Study of mineral fillers in this binding system should be complex to forecasting the final properties of road materials as well as development of application guideline.

Acknowledgement

This research work is financially supported by Ministry of Education and Science of Russian Federation in framework of State Assignment №11.9329.2017/BCh, Russian Federation Ministry of Education and Science in the framework of the Presidential Scholarship № PS-2099.2015.1 as well as Grant № 15-33-51017, using equipment of High Technology Center at BSTU named after V.G. Shoukhov.

References

[1] Putilin E I and Tsvetkov V S 2003 *Application of fly ashes and bottom-ash mixtures when road construction: review of domestic and overseas experience of using of solid-fuel combustion wastes from power stations* (Moscow: Sousdornii)

[2] Zimmer F V 1970 Fly Ash as a Bituminous Filler *Proc. of the 2nd Ash. Utilization Symp.* pp 49–76

[3] Simms S A 1998 *Use of coal fly ash in asphalt concrete mixes* (Halifax: DalTech-Dalhousie University)

[4] Solovyov L A 2004 Includes Ritvield and Derivative Difference Minimization (DDM) methods *J. of Applied Crystallografy* vol **37** pp 743–749.

[5] ASTM C618 2003 *Standard Spec. Coal Fly Ash and Raw or Calcined Natural Pozzolan*

[6] Dvorkin L I 2007 *Construction materials based on industrial wastes: workbook* (Rostov-on-Don: «Fenix»)

[7] Kozhuhova N I, Zhernovskiy I V, Osadchaya M S, Strokova V V and Tchizhov R V 2014 On the Question of the Choice of Natural and Man-Made Materials for Geo-Polymer Binders *Research J. of Appl. Sci.* vol **9(12)** pp 1034–39

[8] Russian Standard 52129–2003 2004 *Mineral powders for asphaltic concrete and organomineral mixtures. Specifications* (Moscow: Russian Gosstroy,)

[9] Kuznetsov D A, Agamyan B S and Baranov T R 2013 *Bulletin of Belgorod State Technol. University named after V.G. Shoukhov* vol **6** pp 43–45

[10] Vyisotskaya M A and Fedorov M Y 2013 *Bulletin of Belgorod State Technol. University named after V.G. Shoukhov* vol **6** pp 61–65

[11] Gezentsvey L B, Gorelyishev A M, Boguslavsky I V and Korolev I V 1985 *Road asphalt concrete* (Moscow: Transport)

IOP Conf. Series: Materials Science and Engineering **222** (2017) 012006 doi:10.1088/1757-899X/222/1/012006

Toward zero waste to landfill: an effective method for recycling zeolite waste from refinery industry

K Homchuen[1], R Anuwattana[2], N Limphitakphong[3] and O Chavalparit [1,3]

[1]Department of Environmental Engineering, Faculty of Engineering, Chulalongkorn University, Bangkok, 10330 Thailand
[2]Environment and Resources Department, Thailand Institute of Scientific and Technological Research, Pathumthani, 12120 Thailand
[3]Research Unit of Environmental Management and Sustainable Industry, Faculty of Engineering, Chulalongkorn University, Bangkok, 10330 Thailand

Email: orathai.c@chula.ac.th

Abstract. One-third of landfill waste of refinery plant in Thailand was spent chloride zeolite, which wastes a huge of land, cost and time for handling. Toward zero waste to landfill, this study was aimed at determining an effective method for recycling zeolite waste by comparing the chemical process with the electrochemical process. To investigate the optimum conditions of both processes, concentration of chemical solution and reaction time were carried out for the former, while the latter varied in term of current density, initial pH of water, and reaction time. The results stated that regenerating zeolite waste from refinery industry in Thailand should be done through the chemical process with alkaline solution because it provided the best chloride adsorption efficiency with cost the least. A successful recycling will be beneficial not only in reducing the amount of landfill waste but also in reducing material and disposal costs and consumption of natural resources as well.

1. Introduction

Most of refinery wastes are difficult to manage as it contaminates with metal and oil [1]. It is classified as hazardous waste which is usually used just once and disposed to secure landfill directly even some of them might be reusable or recyclable. Among several types of landfill waste from refinery industry, one-third of them is spent chloride zeolite generated in gas separation unit for removing hydrogen chloride (HCl) gas from hydrogen stream since it can combine with ammonia to form an ammonium chloride salt resulting in corrosion and fouling or pressure drop in the process consequently [2].

To improve the current waste management system toward zero waste to landfill, sustainable waste management such as 3Rs (reduce, reuse, recycle) strategy should be implemented. Some researchers have used sodium chloride (NaCl) or sodium hydroxide (NaOH) solutions to regenerate zeolites and state that zeolites can nearly be completely regenerated at pH 11-12, while some have been introduced for regenerating spent zeolite via electrochemical method because of its less adsorbent loss, in-situ regeneration and, high regeneration efficiency [3,4,5,6,7].

This study, therefore, aimed to determine the suitable method for recycling zeolite waste from refinery industry in Thailand. Two regeneration methods (chemical and electrochemical process) were examined and compared to find the most effective way. Properties of fresh, spent, and regenerated zeolites were tested. On each experiment, optimal conditions were measured, and an efficiency of chloride adsorption, a residual chloride in regenerated zeolite, and an operation cost were investigated.

2. Materials and method

2.1. Materials

- Fresh and spent chloride zeolites obtained from refinery plant in Rayong Province.
- An electrochemical reactor with dimension of 5 cm x 1 cm x 10 cm was set up by placing flat and rectangular Ti/Pt electrodes with surface area of 25 cm2 in parallel at a separation distance of 1 cm and putting 10 grams of zeolite waste in the middle.
- Chemicals; Sulfuric acid (H_2SO_4), Nitric acid (HNO_3) and Sodium hydroxide (NaOH), were used for adjustment the pH of water and in the regeneration of chemical process.

2.2. Experimental design and analytical method

The experiment of chemical process was designed by varying a concentration of chemical and reaction time, whilst electrochemical process was stimulated for understanding a range of current density, initial pH and reaction time. Both experiments were operated at room temperature of 28 °C.

The characteristics of all zeolites were analysed by X-ray fluorescence spectroscopy (XRF). The surface area, pore size and pore volume were measured via Brunauer-Emmett-Teller method (BET). The morphology was analysed by scanning electron microscopy (SEM). The crystalline structure was tested by X-ray diffraction Spectroscopy (XRD) and the concentration of chloride was determined according to EPA method.

3. Results and discussion

3.1. Properties of zeolite before and after regenerating

To ensure a probability of recycling zeolite, physicochemical characteristics and crystalline structure of the zeolite waste were examined before regenerating to compare with the fresh one. It was found that chemical compositions of both substances were similar, except the composition of chloride as zeolite waste would contain a higher percentage of chloride after spending in HCl removal process as shown in table 1. In addition, the structure of each zeolite in figure 1 was in a crystal form with size of 2 - 4.5 μm. The photograph of zeolite regenerated in alkaline solution (figure 1F, 1G, 1I) was very similar with the fresh one. Thus, it might be possible for recycling zeolite waste instead of landfilling the fresh one once used for removing HCl.

3.2. Optimum conditions of zeolite regeneration via chemical process

In total, 69 experiments were run to examine the effect of each parameter on the substances. Of that, 25 operations were set up by varying a concentration of chemicals from 0.1 to 3 molar (M), and the reminder was for testing a reaction time from 10 to 150 minutes (min). As demonstrated in figure 2, regenerating with acid solution (H_2SO_4 and HNO_3) provided a higher chloride desorption efficiency than that using alkaline solution (NaOH).

An increasing of chemical's concentration affected a decreasing of chloride residue in regenerated zeolite, which meant chloride desorption efficiency increased. However, pH of the solutions will increase as well because sodium-chloride in spent zeolite was released [8]. The best option for regenerating zeolite was found at 0.3M for all three solutions, then such concentration was used to study the effect of reaction time consequently. The optimal reaction time of regeneration with 3M of H_2SO_4 was 50 min with chloride residual in regenerated zeolite of 1.91 mg chloride/g zeolite, while regenerating with HNO_3 and NaOH required 60 min and remained more chloride residual in regenerated zeolite. However, the crystalline structure after regenerating emphasized that spent zeolite regenerated in alkaline condition looked very likely to conform to the fresh zeolite, especially the crystalline structure of zeolite regenerated with NaOH (figure 2-lower)

IOP Conf. Series: Materials Science and Engineering **222** (2017) 012006 doi:10.1088/1757-899X/222/1/012006

Table 1. Physicochemical characteristics of zeolites before and after regenerating.

Substance	**Chemical composition** (wt.%)						**Surface area**	**Pore volume**	**Pore size**
	Cl	SiO_2	Al_2O_3	Na_2O	MgO	SO_3	(m^2/g)	(cc/g)	Å
Before Regenerating									
Fresh zeolite	0.16	46.1	25.7	15.2	2.3	0.85	517	0.421	32.54
Zeolite waste	2.13	43.8	27.3	13.6	2.0	0.84	436	0.379	34.80
After Regenerating via chemical method									
H_2SO_4 0.3 M	0.18	40.7	15.7	6.63	2.0	1.79	354	0.304	31.52
HNO_3 0.3 M	0.25	41.5	20.4	9.54	2.7	-	383	0.425	35.82
NaOH 0.3 M	0.40	42.9	25.3	14.1	2.1	-	435	0.418	34.87
After Regenerating via electrochemical method									
pH 1	0.21	39.7	20.3	8.21	1.83	1.81	390	0.322	34.11
pH 5	0.39	40.3	22.1	9.25	1.89	1.67	394	0.323	36.52
pH 8.1	0.54	41.9	25.1	11.8	1.96	0.11	415	0.396	35.01
pH 10	0.47	41.5	24.9	12.5	1.94	-	430	0.415	34.82
pH 13.6	0.43	40.6	24.1	13.2	2.00	-	431	0.418	34.83

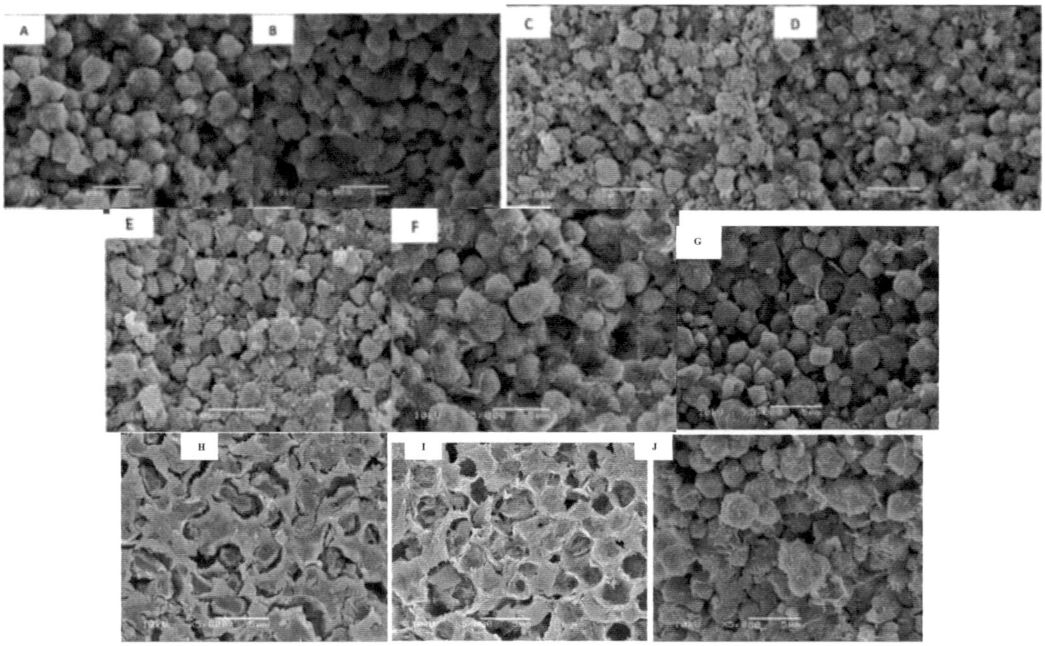

Figure 1. Crystalline structure of zeolites before and after regenerating. (A) Fresh zeolite (B) Spent zeolite (C-G) Regenerated zeolite via electrochemical process at pH 1,5,8.1,10,13.6 (H-J) Regenerated zeolite via chemical process with H2SO4, HNO3 and NaOH

IOP Conf. Series: Materials Science and Engineering **222** (2017) 012006 doi:10.1088/1757-899X/222/1/012006

In conclusion, the optimum condition of chemical process was regenerating zeolite waste with NaOH at concentration of 0.3 M for 60 min. With such condition, the amount of residual chloride in regenerated zeolite was 3.86 mg chloride/g zeolite with chloride adsorption efficiency of 82%.

3.3. Optimum conditions of zeolite regeneration via electrochemical process

In total, 32 electrochemical batches were run to examine the effect of each parameter on the substances. Of that, 13 operations were set up by varying a current density from 4 to 100 mA/cm^2, 10 batches were built for testing an initial pH in a range of 1 – 13.6, and the reminder was for testing a reaction time varied from 0 to 40 min. As demonstrated in figure 3, spent zeolite regenerated in alkaline condition looked very likely to conform to the fresh zeolite, especially the crystalline structure of zeolite regenerated at initial pH of 10 (figure 3A) and the lowest point of electrical resistance (50 ohm) was given at the current density of 20 mA/cm^2 and it was a turning point for either chloride residual in regenerated zeolite or chloride desorption efficiency as well (figure 3B, 3C). Using acid solution took less reaction time than operating on neutral or alkaline solution (figure 3D) and the results agree with Liu et al (2008) and Elsener and Angst (2007) that indicated an increasing of chloride removal efficiency or a decreasing of chloride residue in regenerated zeolite with the reaction time increased.

To sum up, the optimum condition of electrochemical process was regenerating zeolite waste with Ti/Pt electrodes at initial pH of 10 and current density of 20 mA/cm^2 for 25 min. With such condition, the amount of residual chloride in regenerated zeolite was 4.89 mg chloride/g zeolite with chloride adsorption efficiency of 77%.

3.4. Cost of operation

Referring to an interview with project manager of zero waste to landfill for refinery plant, one-third of total waste to secure landfill was zeolite. A million of note was paid for handling such waste properly annually due to a high cost of secure landfill of 3,200 THB/kg of waste. In this study, cost of regenerating via chemical process was accounted only in term of chemical used, while electrochemical process involved cost of chemical and power. It was found that for every kilogram of zeolite regenerated, only 10 THB was paid for regenerating zeolite waste with NaOH solution at the optimum conditions. Whilst to reach the optimum conditions of electrochemical process, around 160 THB/ kg of zeolite regenerated was required mainly for electricity bill.

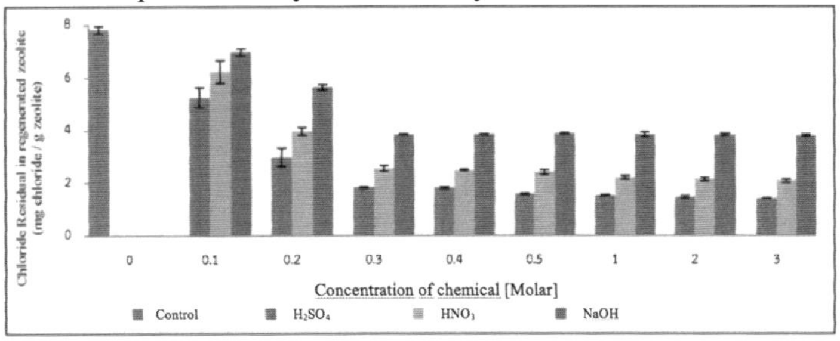

(a)

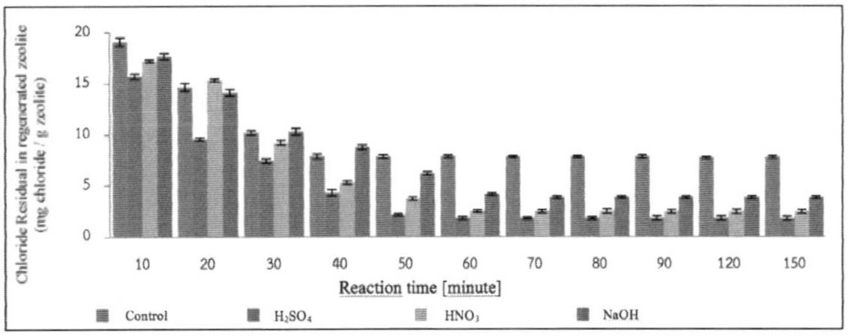

(b)

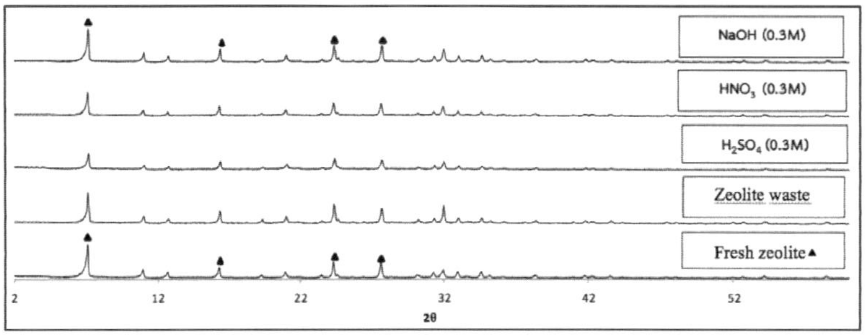

(c)

Figure 2. Optimum conditions of regenerating zeolite waste via chemical process. (Upper) Effect of chemical concentration on chemical regeneration (Middle) Effect of reaction time on chemical regeneration (Lower) Crystalline structure after regeneration

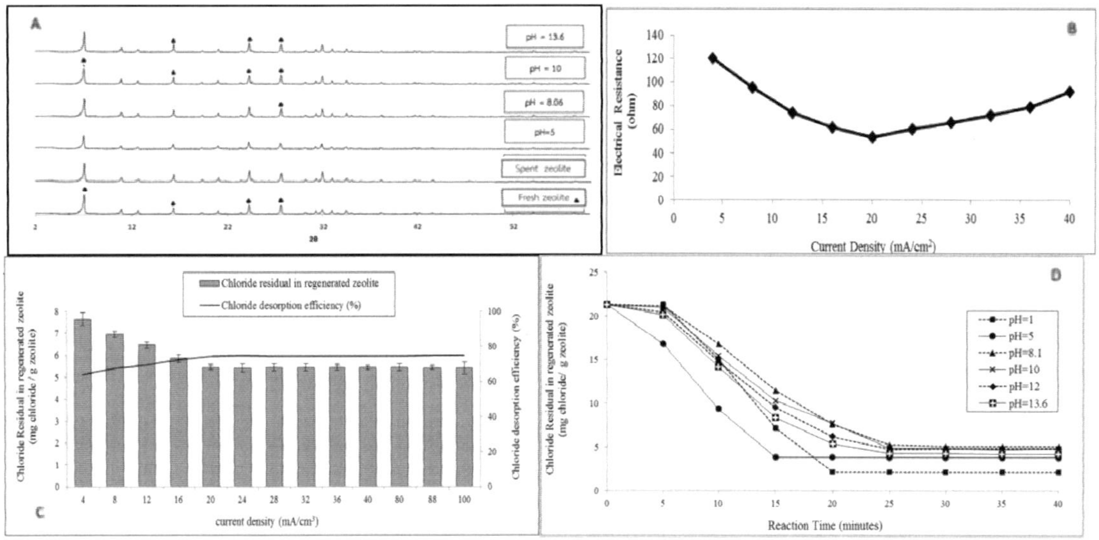

Figure 3. Optimum conditions of regenerating zeolite waste via electrochemical process. (A) Crystalline structure after regeneration (B) Effect of current density on electrical resistance (C) Effect of current density on electrochemical regeneration (D) Effect of reaction time on electrochemical regeneration

4. Summary

To improve waste management system toward zero waste to landfill, sustainable waste management such as 3Rs strategy was needed to determine a suitable method for recycling zeolite waste from refinery industry in Thailand. Chemical and electrochemical processes were introduced for regenerating the spent zeolite instead of using a fresh zeolite once as usual. The optimum conditions of both processes were examined and the chloride adsorption efficiency and cost of operation at the optimum conditions were compared to identify the most effective method of regeneration. From all point of view; chloride adsorption efficiency, residual chloride in regenerated zeolite and operation cost, regenerating zeolite waste via chemical process at the optimum conditions was more effective than that of electrochemical process. Therefore, chemical process is an alternative effective option for regenerating zeolite waste from refinery industry. Moreover, the regeneration process has shown several advantages such as a reduction of waste, time and cost for landfilling, a reduction of natural resources for producing a fresh zeolite, and a saving cost of purchasing a fresh zeolite. The further study might be focused on an application of zeolite regenerated with sulfuric acid solution because the higher sulfur residual in regenerated zeolite might be beneficial for toxic or hydrogen sulfide adsorption.

References

[1] Usapein P and Chavalparit O 2014 *Mater Cycles Waste Manag* vol 32 p 509
[2] Information on http://www. digitalrefining.com/data/articles /file/433058853.pdf
[3] Li M, Feng C, Zhang Z, Lei X, Chen N and Sugiura N 2010 *Micropor Mesopor* vol 127 p 161
[4] Lei X, Li M, Zhang Z, Feng C, Bai W and Sugiurac N 2009 *J. Hazard. Mater.* vol 169 p 746
[5] Alejandro S, Valdés H and Manéro M 2014 *J. Hazard. Mater.* p 212
[6] Eduard S. and Călinescu I 2014 *U.P.B. Sci. Bull* vol 76 p 96
[7] Hussain S N, Asghar H M A, Sattar H, Brown N W and Roberts E P L 2015 *J. Ind. Eng. Chem.* vol 30 p 212
[8] Piwgaew W, Grisdanurak N and Kumsaen T 2012 4th *Int. Conf of KKU* (Khon Kaen, Thailand)

Acknowledgement

This research was funded by the Ratchadapisek Sompoch Endowment Fund (2016), Chulalongkorn University (CU-59-002-IC).

IOP Conf. Series: Materials Science and Engineering **222** (2017) 012007 doi:10.1088/1757-899X/222/1/012007

Effect of solution molarity on properties of hydrothermally grown ZnO nanostructures

Yi Chen[1], Wei-You Chen[1], Hung-Pin Hsu[1], Jiunn-Chyi Lee[2] and Ya-Fen Wu[1,3*]

[1]Department of Electronic Engineering, Ming Chi University of Technology, New Taipei City 243, Taiwan
[2]Department of Electrical Engineering, Taipei City University of Science and Technology, Taipei 112, Taiwan
[3]College of Engineering, Chang Gung University, Taoyuan 333, Taiwan

Email: yfwu@mail.mcut.edu.tw

Abstract. In this work, the hydrothermally grown ZnO nanostructures were successfully prepared on indium tin oxide glass using different aqueous solution molarities. The effect of solution molarity on the structual and optical properties of ZnO nanostructures were studied by X-ray diffraction and temperature-dependent photoluminescence (PL) measurements. The PL intensity ratio of the ultraviolet emission to the visible emission is also calculated. It is found that the crystal quality and the optical properties are improved by raising the solution molarity, but too higher solution molarity should lead to the disorder of ZnO structures, and deteriorate the crystal quality.

1. Introduction

Zinc oxide (ZnO) has been extensively investigated due to its unique properties, including wide direct bandgap energy (3.36 eV), large exciton binding energy (60 meV), high piezoelectricity, high electron mobility, and thermal stability. These characteristics make ZnO nanostructures interesting for application in ultraviolet (UV) light emitters, optoelectronic devices, and chemical sensors as well as acousto-optic, electro-optic, and piezoelectric material [1-3]. ZnO nanostructures with small dimensions exhibit distinctive properties arising from their nanoscale size and shapes. The roles of the nanostructure size, impurities, and morphology are very important to their applications. ZnO nanostructures have been synthesized by different methods, including metal-organic chemical vapor deposition (MOCVD) [4], physical vapor phase deposition [5], and the low temperature hydrothermal method [6]. The hydrothermal method, which involves heterogeneous nucleation in super saturated solutions for the growth of nanostructures on the surface, has been widely used to ZnO structures on ZnO-seeded substrate. The advantages of this method include low-cost, low-temperature, nanoelectronic compatible, and suitable for large area substrates. The properties of ZnO produced by the hydrothermal method are dependent upon preparation parameters such as growth temperature, solution molarity, and deposition time [7]. In this work, the hydrothermally grown ZnO nanostructures were prepared with different aqueous solution molarities. Measurements of X-ray diffraction (XRD) and temperature dependent photoluminescence (PL) were performed to examine the structural and optical properties of the ZnO nanostructures.

2. Experimenal

A thin film of zinc acetate (comprised of a solution containing zinc acetate dihydrate $Zn(CH_3COO)_2{\cdot}2H_2O$) was spin coated onto an indium-tin-oxide (ITO) glass substrate and annealed at 300 °C in oxygen for 20 min. The hydrothermal method was used to grow ZnO nanostructures on ITO glass using an aqueous solution of zinc nitrate dehydrate ($Zn(NO_3)_2$) and hexamethylenetetramine (HMTA) which were dissolved in 100 ml of deionized (DI) water. The molarities of solutions were set to 10, 20, 30, 40, and 50 mM and the growth temperatures were set to 90 °C. The growth was carried out for 3 hours after the temperature of the aqueous solution reached 90 °C. Finally, the sample was taken out from the aqueous solution and immersed into DI water to remove any residue on the substrate. After growth, the samples were characterized by HRXRD. Rocking curves of the (002) reflection were produced using the Bede D1 system. The temperature dependent PL measurements were carried out by mounting the sample in a closed cycle helium cryostat where the temperature (T) was varied from 10 to 300 K and using a He-Cd laser operating at a wavelength of 325 nm and the average excitation intensity was 30 mW. The luminescence signal was dispersed through a 0.5 m monochromator and detected using a silicon photodiode by standard lock-in amplification technique.

3. Results and discussion

Figure 1 shows the XRD patterns of the ZnO nanostructures with the solution molarities from 10 to 50 mM. The nanocrystalline quality could be analyzed using XRD measurements. It can be seen that the detected diffraction peaks from the nanocrystallite responses have a wurtzite hexagonal ZnO structure and the narrow broadenings indicate good crystallinity of the as-synthesized products. The distinct diffraction of the ZnO can be indexed as corresponding to the crystal planes of ZnO (100), ZnO (002), ZnO (101), ZnO (102), and ZnO (110), respectively. With increasing solution molarities, the intensity of XRD peaks increases. However, the full width at half maximum (FWHM) of the XRD spectra decreases as the solution molarity increases from 10 to 40 mM, and then increases as the solution molarity further increases to 50 mM. The widest FWHMs of the XRD peaks in the highest-molarity ZnO nanostructures imply that this sample has a lower crystalline quality than the other samples. With high solution molarity during hydrothermal process, the growth rate may proceed too fast for regular morphology and enhance the degree of the overall nanocrystallite size fluctuation, inducing broadened FWHMs of XRD spectra [8].

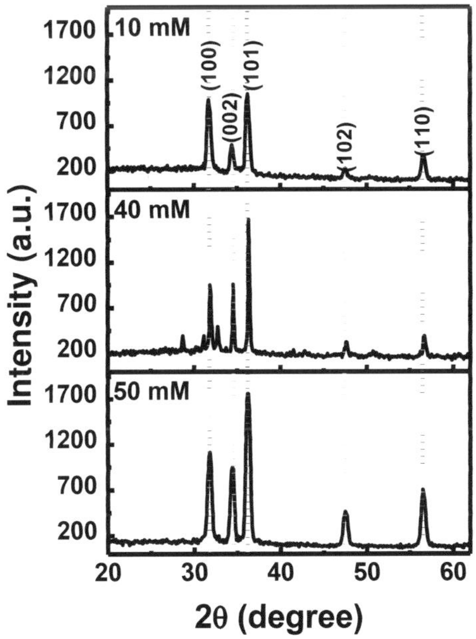

Figure 1. High-resolution X-ray diffraction spectra for the hydrothermally grown ZnO nanostructures with solution molarity 10, 40 and 50 mM, respectively.

Figure 2 shows the PL spectra of the ZnO nanostructures synthesized using different solution molarities at T=300 K. The ZnO nanostructures exhibit weak emission in the ultraviolet (UV) region at approximately ~3.2 eV (see inset of Figure 2) and a strong and broad yellow-orange emission peak at ~2.02 eV. The ultraviolet emission is attributed to free-exciton recombination corresponding to the near-band-edge transition in ZnO [9]. The origin of the emission peak in the visible region is related to the radial recombination of photon-generated holes with single ionized charge of the specific defects such as O vacancies or Zn interstitials [10,11]. Moreover, increasing the solution molarity from 10 to 30 mM leads to a rising of the invisible emission peak intensity. We suppose that the increase in emission intensity could be associated with the increasing concentration of the oxygen vacancies.

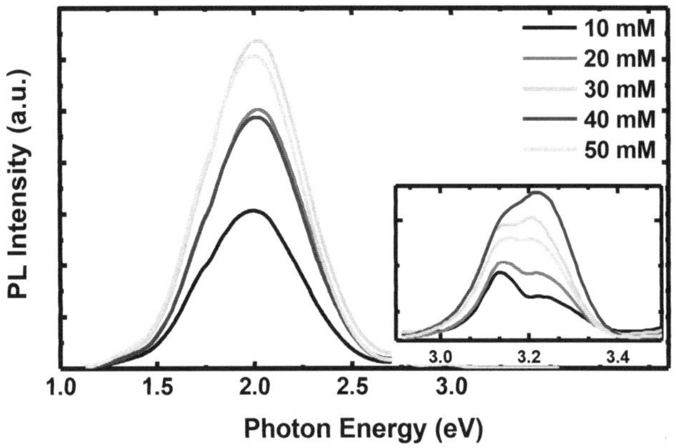

Figure 2. Room-temperature PL spectra of ZnO nanostructures with solution molarities from 10 to 50 mM. The inset shows the magnified UV emission spectral region.

More detailed data on the temperature dependence of the PL peak position and the PL linewidth taken over a broader temperature range are plotted in Figure 3. Figure 3(a) demonstrates the temperature dependent PL emission energy of ZnO sample with solution molarity 30 mM. It is apparent from the plot that the visible yellow-orange emission peaks show anomalous temperature-dependent behaviors, which differ remarkably from the monotonous band-gap shrinkage with increasing temperature. This phenomena could be associated with the defect-related levels in the ZnO nanostructures, and is attributed to the temperature related transitions between the conduction band, shallow donors (such as intrinsic Zn_i), and ionized oxygen vacancy levels [12]. The temperature-dependent FWHMs of PL spectra measured from this sample are exhibited in Figure 3(b). As temperature is increasing, the broadening of PL linewidth is obvious, resulting from the electron-phonon scattering mechanism.

Figure 4 presented the PL intensity ratio of the emission in the UV region (I_{UV}) to the emission in the visible region (I_{vis}). As comparing the optical properties of samples, the value of I_{UV}/I_{vis} of the ZnO nanostructures is an important factor [13]. It can be seen that the ratio increases as the molarity increases from 10 to 40 mM, and then decreases when the molarity further increases to 50 mM. It is known that there are more defects that could act as nonradiative recombination centers and reduce light emission on the surface compared to the internal parts of materials. The higher PL intensity ratio means fewer structural defects in the samples. Therefore, the ZnO nanostructures with molarity 40 mM have a better relative crystalline quality. Too higher solution molarity leads to the disorder of ZnO structures, and deteriorate the crystal quality.

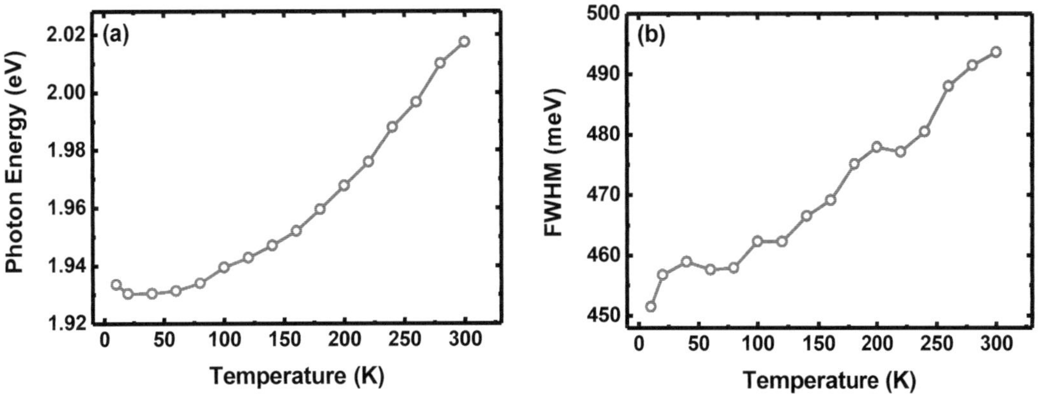

Figure 3. Temperature dependence of (a) peak position and (b) FWHM of PL spectra from the ZnO nanostructures synthesized using solution molarity 30 mM.

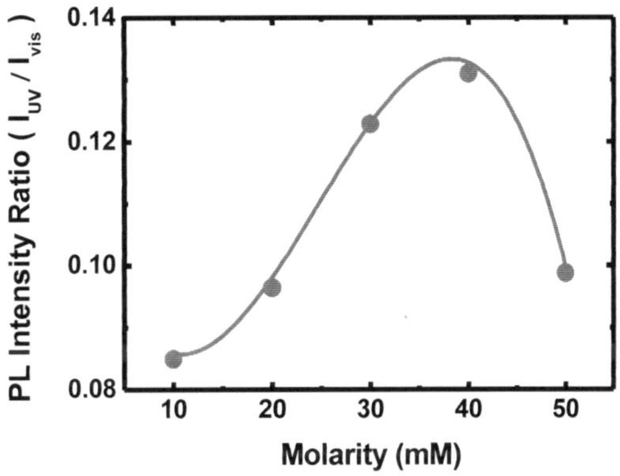

Figure 4. Variation of the PL intensity ratio of the emission in the UV region (I_{UV}) to the emission in the visible region (I_{vis}) values with various solution molarities from 10 to 50 mM.

4. Conclusions

In this paper, the effects of solution molarity on hydrothermally grown ZnO nanostructures were studied. The structural properties of the prepared ZnO nanostructures were determined by XRD measurements, and the optical properties were investigated by the measurement of temperature dependent PL spectra. It is found that the crystal quality and the optical properties are dependent on the solution molarity during hydrothermal process. The intensity of XRD peaks increase and the PL intensity ratio of the UV emission to the visible emission is also raised up with increasing solution molarities. However, the FWHM of the XRD of the ZnO nanostructures become wider and the PL emission intensity ratio (I_{UV}/I_{vis}) degrades with further increase of molarity. Therefore, the structural and optical properties of ZnO nanostructures can be improved by optimizing of solution molarity.

References

[1] Ferro R, Rodríguez J A and Bertrand P 2008 *Thin Solid Films* **516** 2225
[2] Rai P, Song H M, Kim Y S, Song M K, Oh P R, Yoon J M and Yu Y T 2012 *Mater. Lett.* **68** 90
[3] Rouhi J, Alimanesh M, Dalvand R, Ooi C, Mahmud S and Mahmood M R 2014 *Ceram. Int.* **40** 11193
[4] Jeong S H and Aydil E S 2009 *J. Cryst. Growth* **311** 4188
[5] Willander M, Nur O, Sadaf J R, Qadir M I, Zaman S, Zainelabdin A, Bano N and Hussain I 2010 *Materials* **3** 2643

IOP Conf. Series: Materials Science and Engineering **222** (2017) 012007 doi:10.1088/1757-899X/222/1/012007

[6] Al-lami S and Jaber H 2014 *Chem. Mater. Res.* **6** 101
[7] Pál E, Hornok V, Oszkó A and Dékány I 2009 *Colloids Surf. A: Physicochem. Eng. Aspects* **340** 1
[8] Zhao X, Lee J Y, Kim C R, Heo J, Shin C M, Leem J Y, Ryu H, Chang J H, Lee H C, Jung W G, Son C S, Shin B C, Lee W J, Tan S T, Zhao J and Sun X *Physica E* 2009 **41** 1423
[9] Sui C, Lu Z and Xu T 2013 *Optical Materials* **35** 2649
[10] Aguilar C A, Haight R, Mavrokefalos A, Korgel B A Chen S 2009 *ACS Nano* **3** 3057
[11] Xing G Z, Yao B, Cong C X, Yang T, Xie Y P, Li B H Shen D Z 2008 *J. Alloys Compd.* **457** 36
[12] Cao B, Cai W and Zeng H 2006 *Appl. Phys. Lett.* **88** 161101
[13] Jayah N A, Yahaya H, Mahmood M R, Terasako T, Yasui K and Hashim A M 2015 *Nanoscale Res. Lett.* **10** 7

IOP Conf. Series: Materials Science and Engineering **222** (2017) 012008 doi:10.1088/1757-899X/222/1/012008

Improving the quality of biopolymer (poly lactic acid) with the addition of bentonite as filler

Suryani[1,2,a], Harry Agusnar[1], Basuki Wirjosentono[1], Teuku Rihayat[2] and Nurhanifa[2]

[1]Department of Chemistry, Faculty of Mathematics and Natural Science, Universitas Sumatera Utara, 20155, Medan, Indonesia
[2]Departement of Chemical Engineering, Politeknik Negeri Lhokseumawe, 24301, Aceh Indonesia

Email: [a]suryanisalim@yahoo.com

Abstract. PLA (Poly Lactid Acid) - Bentonite polymer nanocomposite which is a combination of natural and nanometer-scale inorganic substances created through three processes, mixing using a melt blending, molding with a hot press using specimens Standard ASTM D 638 Type IV and drying. In this study, PLA combined with two types of natural bentonite obtained from different areas to find differences in the quality of the results of characterization. To optimize the performance of filler, before mixing, bentonite have to furificate first with (NaPO3)6 and also open the interlayer space with CTAB. D-spacing of bentonite imterlayer were analyze by X-Ray difraction (XRD). Characterization bionanocomposite resulting morphologic structure was tested using a Transmission Electron Microscope (TEM). Mechanical analysis of PLA-bentonite nanocomposite in the form of tensile strength was tested using a tensile test specimens of standard American Standard for Testing Materials (ASTM) D 638 Type 4, and thermal resistance using Thermo Gravimetric Analysis (TGA).

1. Introduction

Some material polymer derived from environmentally friendly materials (biopolymers) has generated a lot of Poly Hydroxy alkanoic eg, PHA (PHA), poly hydroxy butyrate (PHB), and Poly Lactid Acid (PLA). However, one of the emerging biopolymers examined are PLA (Poly Lactid Acid) due biodegradation capabilities so high that classified as environmentally friendly polymers. Unlike synthetic polymers are generally derived from petroleum materials, PLA is derived from renewable sources, namely from a variety of plants that have a high starch content such as cassava, sweet potato, banana, corn and so on [1].

IOP Conf. Series: Materials Science and Engineering **222** (2017) 012008 doi:10.1088/1757-899X/222/1/012008

PLA can be formed through a process of esterification of lactic acid obtained by fermentation by bacteria using substrates starches or simple sugars. There are two optical isomers of lactic acid as the main carbon atom binds to different groups, namely poly (L (+) - laktid) (PLLA), poly (D (-) -laktid) (PDLA), and poly (DL-laktid) (PDLLA). PLLA and PDLA usually crystalline because it has the isotactic optically active isomer, while the relative PDLLA amorphous shape because it has the inactive optical isomer atactic. Commercial products are usually in the form of DL-. The process in the synthesis of poly lactic acid according Averous (2008) can be through three methods, namely Polycondensation Direct (direct condensation polymerization), azeotropic dehydration condensation, and Methods Ring Opening Polymerization (ROP). Compared with other methods, the ROP method is a method that is extremely complex and produces PLA with good characteristics for various applications. Ring opening polymerization method that produces PLA with a molecular weight of 2 x 104 to 6.8 x 105 has been patented by Cargill (USA) in 1992 [2].

Another advantage of the PLA include Biodegradable, biocompatible, 100% recyclable and can be burned perfectly and produces CO2 gas and water. However, these biopolymers also has drawbacks one of which is a low melting point so that the heat resistance is not so good. Therefore, to overcome these weaknesses, PLA its properties can be improved by the addition of fillers (filler) to form nano-sized nanocomposite [3]

Bentonite is a mineral's where its use as a filler material economical to modify the creation and performance of the material. In mineralogy bentonite by Gilson in 1960 was defined as a clay that consists of 85% montmorillonite which has the chemical formula (Al2O3.4SiO2xH2O). Bentonite is a mineral resources are abundant in Indonesia. Minerals bentonite has a diameter of less than 2 micrometers are composed of various phyllosilicate containing silica, aluminum oxides and hydroxides which bind water. Bentonite structure consisting of two layers (three layers) composed of two layers of silica tetrahedral and octahedral central one. tetrahedral layers containing silicon atoms / silicate and octahedral layers made of aluminum or magnesium hydroxide. Generally, this layer has a thickness of about 1 nm and can be widened to 30 nm. Among the octahedral and tetrahedral layers are monovalent or bivalent cations, such as Na +, Ca + and Mg 2+. The purpose of the addition of filler is to improve and enhance the material properties of the polymer to make it more stable, stronger mechanical and chemical as well as more resistant to heat in order to function optimally in the various sectors of human life such as household appliances, automotive, agriculture, healthcare and packaging [4].

In general, nanocomposite polymer formed by dispersing organic or inorganic nanoparticles in the polymer matrix. Nanobentonit have a very large surface area so that it can interact effectively with the polymer matrix at low concentrations (2-8%). Various process for forming a bentonite-polymer nanocomposite, ie in situ polymerization, exfoliation solution and melt intercalation. On learning this, the PLA-bentonite nanocomposite formed through melt intercalation bentonite is mixed directly into the polymer matrix in the molten state. Sodium hexametaphospate (SHMP) or (NaPO3) 6 is used as deffloculant to purify bentonite before mixed with polymers with the separation of montmorillonite with other substances that are impurities and disturbing keoptimalan interact with the filler in the polymer matrix. CTAB (cetyl trimethyl ammonium bromide) is used to unlock the layer (layer) in order to more bentonite expands so as to maximize the intercalation of bentonite in the polymer matrix. After purification, the content in the bentonite was investigated using X-ray apparatus Difraction (XRD) while seeing an increasing number of montmorillonite contained [4].

PLA-bentonite nanocomposite formed tested strength using a tensile test specimens of standard ASTM D 638 Type 4. Then morphologic structure was analyzed under the tool Transmission Electron Microscope (TEM). To determine the increase of the polymers to heat, PLA-bentonite nanocomposite analyzed using Thermo Gravimetric Analyzer (TGA).

2. Experimental

2.1. Material

PLA (Poly Lactid Acid) crystalline type used in the study came from Nature Works Co. (USA), while bentonite used in the study came from two different regions of the region, namely bentonite I Nisam, Aceh Utara (UA's Bentonite) and Bentonite II of the District. Bener Meriah, Takengon, Central Aceh (Tkg's Bentonite). Other additives that are used in the form of distilled water, cetyl trimethyl ammonium bromide (CTAB) and sodium hexametaphosphate $(NaPO_3)_6$ for the purification of bentonite.

2.2. Methods

2.2.1. Purification of bentonitetop of form

Prepared two containers for purification of two types of natural bentonite. Each Bentonite weighed as much as 3 grams are weighed and then in grinding for 60 minutes. After sifting it done at a size of 230 mesh. Bentonite grinding results put into 250 ml of distilled water and dispersed by mixing for 2 hours on a magnetic mixer until it forms a suspension. Then added 1,86 grams of CTAB and also (NaPO3) 6 as much as 0.06 grams and carried back in dispersing magnetic mixer for 6 hours at a speed of 50 rpm. Results disenstrifugasi dispersion for 2 minutes at a speed of 700 rpm and is separated from the liquid. Results centrifuging and then washed with distilled water to remove any remaining back (NaPO3) 6 overproduction. Dried in an oven with a temperature of 60 ° C for 1 day [5].

2.2.2. Nanocomposites preparation

Prepared 3 PLA samples each of 20 g. Bentonite is mixed filler PLA and I purified beforehand by comparison, namely: 20 g: 0.2 g (1% by weight polymer); 20 g: 0.6 g (3% by weight of polymer) and 20 g: 1.0 g (5% by weight of polymer). Melt blending the mixture melted at the temperature of 140 ^{o}C, then hardens by itself to form PLA-Bentonite Nanocomposite. PLA-bentonite nanocomposite further cut into granules and inserted into a mold specimens according to Standard ASTM D 638 Type IV, then compressed with a hot press at a temperature of 190 oC. PLA-bentonite nanocomposite that has been printed is then dried in a vacuum oven with a temperature of 60 ° C for 24 hours or can be dried in ambient air [6].

2.3. Characterization

The extent of dispersion of the bentonit in the filled samples and also the nanobentonite interlayer spacing was examine by X-Ray Difraction (XRD) which performed at room temperature using a Shimadzu XRD 600 X-Ray diffractometer (30 kV, 30 mA) with a nickel CuKα radiation source. The data was colected over a range of scattering angels (2θ) of (1-10°) [7].

In order to know the tensile properties, the tensile strength test were carried out with an Instron 4468 machine according to ASTM D 638 Type IV. Samples were cut to 165 x 19 x 0,5 mm in size and straining was carried out at 50 mm/min crosshead speed. For each spesiment, five samples were tasted and average values of the tensile strength was calculated.

The morphology of the disk-shaped nanocomposites was investigated by Transmission Electron Microscope (TEM). The TEM samples were cut in thin section (50-80 nm thickness) at -100 °C using an ultramicrotome with a diamond knife.

Thermo Gravimetric Analyzer (TGA) used to determine the glass transition temperature (Tg) of PLA-bentonite nanocomposites which carried out using a Perkin-Elmerinstrument TGA equipment at 1 Hz from -98 to 140 °C with a heating rate of 5 °C/min in tensile mode. The sample was cut into just a small size. The position of the maximum tan δ value was used to determine the glass transision temperature (Tg).

3. Result and Discussion

3.1 XRD analysis

XRD was used to examine the extent of dispersion of the bentonit in the filled samples (PLA) and also the nanobentonite interlayer spacing. In Figure 1a the pure AU's bentonite-PLA shows a diffraction peak at $2\theta = 3.870^o$, indicating that the d-spacing between the silicate layer was about 1.163 nm. Meanwhile, the broad difrraction peaks in 1%, 3% and 5% of AU's bentonite were $2\theta = 6.356^o$ (d-spacing = 1.478 nm), 7.137^o (d-spacing=1.953 nm), and 5.901^o (d-spacing = 2.677 nm) respectively. This implies that after the cation exchange reaction in CTAB-montmorillonite was larger than that of the pure bentonite. This is turn indicates that in bentonite was succesfully intercalated in polymer matrix in increase of its content.

Table 1. Summary of the d-spacing and 2θ values for pure bentonit, PLA-AU's bentonite and PLA-Tkg's bentonite

	2θ (degree)	**d-spacing**
Pure AU	2,0	1.163 nm
Pure Tkg	2,0	1.105 nm
1% AU-PLA	2,0	1.478 nm
3% AU-PLA	2,0	1.953 nm
5% AU-PLA	2,0	2.677 nm
1% Tkg-PLA	2,0	1.372 nm
3% Tkg-PLA	2,0	1.731 nm
5% Tkg-PLA	2,0	2.071 nm

Figure 1b also present three series of XRD curves of PLA - Tkg's bentonite nanocomposites. In the case of 1%, a very small basal spacing was observed, from 1.105 nm in pure-bentonite to 1.372 nm. In 3%, the d-spacing layer was 1.371 nm. With increasing bentonite content, the peak became stronger and exhibited a sharply increased peak toward the d-spacing of 2.077 nm for 5% Tkg's bentonit-PLA. Meanwhile, these results suggest that intercalated PLA-Tkg's bentonites nanocomposites was low than the AU's bentonites. It caused the different swallen spacings in these two bentonites which are influenced of CTAB attributes and bentonites composition. Table 1 summaries the d-spacing and 2θ (degrees) for PLA-AU's bentonite and PLA-Tkg's bentonite.

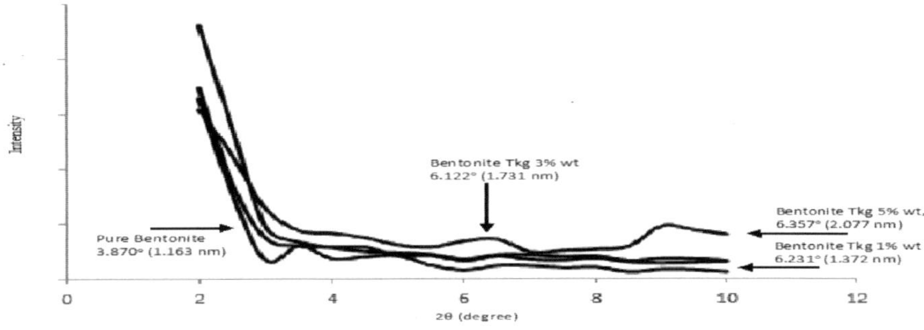

(a)

Figure 1a. The XRD Patterns of pure Aceh Utara's bentonite, PLA-bentonite Aceh (1,3,5 wt %) after treatment

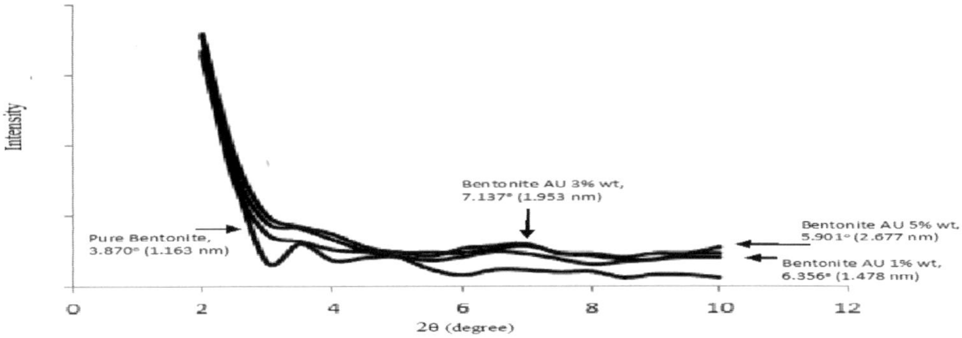

(b)

Figure 1b. The XRD Patterns of pure Takengon's bentonite, PLA-bentonite (1,3,5 wt%) after treatment

3.2 Mechanical properties

The mechanical properties of PLA-bentonite nanocomposites are given in Figure 2. The pure bentonites had a remarkably beneficial effect on the tensile strength of the nanocomposites which inscreased with increasing bentonit content (1,3, and 5%) having highest values in each of bentonite (AU's and Tkg's). Figure 2 explains the tensile strength of pure PLA and PLA-bentonit nanocomposites were usually higher than without.

In the result, PLA which used a mixture of Tkg's bentonite has lower of tensile strength than PLA with AU's bentonite on each % increase. This indicates that the quality of bentonite AU better in improving the tensile strength of polymer. This is due to the content of montmorillonite in the AU's bentonite is much larger than the Tkg has.

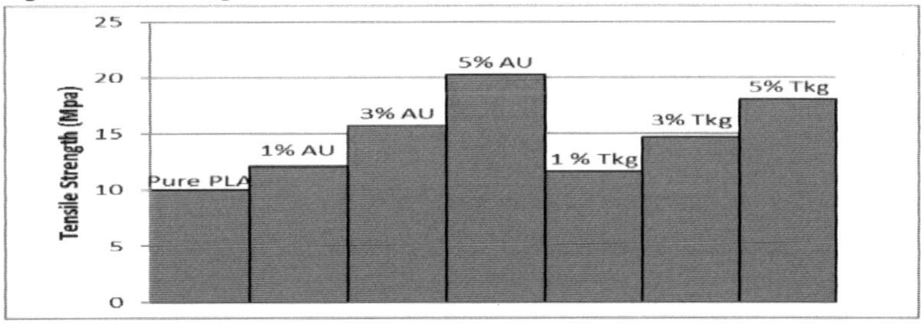

Figure 2. Tensile Strength of pure PLA and PLA-Bentonite nanocomposites

3.3 Morphology analysis

Figure 3 shows the TEM micrographs of PLA-bentonite nanocomposites prepared by 5% of bentonit inserted in polymer matrix which is taken by the most excellent quality. Figure (a) and (b) shows the surface structure of the polymer with bentonite AU's and Tkg's where it can be seen that the surface is not much different.

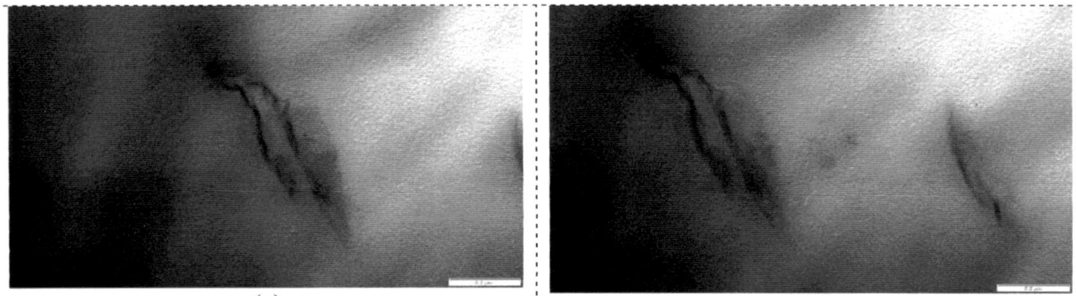

(a) (b)

Figure 3. TEM Micrographs of PLA-bentonit nanocomposites with (a) 5% Aceh Utara's bentonite (b) 5% Takengon's bentonite.

3.4 Thermal stability

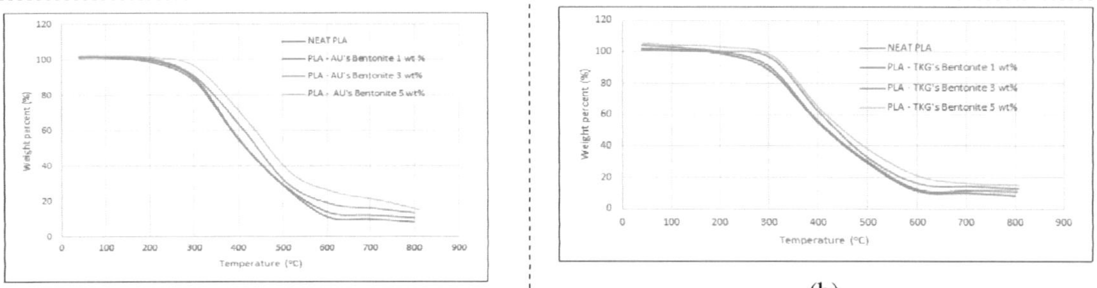

(a) (b)

Figure 4. (a). The TGA curves of Neat PLA and PLA – AU's Bentonite 1, 3, 5 wt% ; (b) The TGA curves of Neat PLA and PLA – TKG's Bentonite 1, 3, 5 wt%

The thermal properties were determined using thermogravimetric analysis (TGA). The TGA analysis of neat PLA, PLA – AU's Bentonite 1, 3, 5 wt% and PLA – Tkg's Bentonite 1, 3, 5 wt%, respectively, are shown in Figure 4 (a) and (b). With regard to neat PLA, the degradation at 200-450 °C is attributed to depolycondensation reaction. Then at higher temperature, the material degrades slowly, and it degrades completely at about 700°C. The results show that as compared to pure PLA, thermal resistance is enhanced in the presence of bentonite. This indicates an improvement in thermal stability of PLA because the organic material can prevent the heat from expanding quickly and limit further degradation. Onset degradation of PLA is at 250°C, and is lower than of the PLA – AU's bentonite and PLA – Tkg's bentonite takes place at about 350°C and 367°C, respectively. It means the thermal resistance of nanocomposites improved above 70% compared pure polymer.

4. Conclusion

The combination of PLA-bentonite as nanocomposite aims to improve the properties of PLA as a polymer. Bentonite was intercalated into the polymer matrix by melting intercalation. The raw bentonite used came from two different regions to look for differences in quality. The d-spacing of bentonite interlayer was the investigation and the result is bentonite from Aceh Utara more swell than bentonite from Takengon. In the analysis of tensile strength, the addition of Aceh Utara's bentonite also shows the results of tensile strength better than Takengon's bentonite, it caused of montmorillonite content and as the increasing of % of bentonite make it more strong. For morphology structure, PLA with the addition of bentonite of Aceh Utara and also Takengon shows the results that were not significantly different that was investigated under TEM.

References

[1] Lopes, M S. (2014), "Synthesis And Charactrization of Poly (Lactid Acid) by Ring Opening Polimerization for Biomedical Application", *Chemical Engineering Transsaction,* Vol.38,333.

[2] Avérous, L. (2008), "Synthesis, Properties and Applications, In Monomers, Polymers and Composites from Renewable Resources", *Poly Lactid Acid*, 21, hal 223-225.

[3] Rola Mansa, Chih-Te Huang, Ana Quintela, Fernando Rocha, Christian Detellier. (2015), "Preparation and characterization of novel clay/PLA nanocomposites", *Applied Clay Science*, 115, 87–96

[4] Anja Terzića, Lato Pezo, Ljubiša Andrić, Vladimir B. Pavlović, Vojislav V. Mitić (2017), " Optimization of bentonite clay mechano-chemical activation using artificial neural network modeling", *Ceramics International*, 43, 2549–2562

[5] Emilie Picard, Eliane Espuche, René Fulchiron, (2011), " Effect of an organo-modified montmorillonite on PLA crystallization and gas barrier properties", *Applied Clay Science,* 53, 58–65

[6] Gong, Z., Liao, L., Lv, G., Wang, X, (2012), "A Simple Method For Physical Purification Of Bentonite", *Applied Clay Science,*119, 294–300.

[7] Najafi, N., Heuzey, M.C., Carreau, P.J, (2012), "Polylactide (Pla)-Clay Nanocomposites Prepared By Melt Compounding In The Presence Of A Chain Extender", *Composites Science and Technology*, 72, 608-615.

[8] Teuku Rihayat, Suryani Salim, Harry Agusnar, Fajri and Zaimahwati (2015), "Synthesis of polyurethane/clay nanocomposites based palm oil polyol coating", *Journal of Mechanical Engineering and Sciences*, 9, pp. 1580-1586

IOP Conf. Series: Materials Science and Engineering **222** (2017) 012009 doi:10.1088/1757-899X/222/1/012009

Anaerobic biodegradation of polylactic acid under mesophilic condition using thermal-alkaline pretreatment

Sutisa Samitthiwetcharong[1,a], Pratin Kullavanijaya[1,b] and Orathai Chavalparit[1,2,c]

[1]Department of Environmental Engineering, Chulalongkorn University, Bangkok 10330, Thailand
[2]Research Unit of Environmental Management and Sustainable Industry, Faculty of Engineering, Chulalongkorn University, Bangkok 10330, Thailand

Email: [a]sutisa.smitth@gmail.com, [b]mail2pratin@gmail.com, [c]orathai.c@chula.ac.th

Abstract. The propose of this study was to investigate the effect of thermal-alkaline pretreatment with emphasis on sodium hydroxide concentration (NaOH), temperature and reaction time, on enhancement of polylactic acid (PLA) films degradation and biogas production. The results found that NaOH concentration and reaction time were two main parameters influencing on PLA degradation. While, less significant was found for temperature. From the Response Surface Methodology (RSM), it was concluded that the optimum pretreatment conditions were at 0.5 M of NaOH, temperature of 60°C and 24 hr of reaction time. This was generated about 3.7 times (215.47 ml/gVS_{added}) higher gas production comparing to non-pretreated PLA films which was 58.28 ml/gVS_{added}. The maximum biodegradability of PLA film was 20.14%. This was estimated to be 4.7 times higher than non-pretreated PLA (4.32%). This finding demonstrated the benefit of thermal-alkaline pretreatment on surface of PLA films destruction. Consequently, the microbial enzymes could degrade PLA more easily, resulting in an increase of biogas production.

1. Introduction

In order to reduce the consumption of petroleum based plastics, bioplastic mainly PLA is widely utilized in many countries. Due to the environmental friendly and renewable resources while its processing, the releases of greenhouse gases is lower than other petroleum based plastics. Moreover, the biodegradable plastics are completely degraded under the appropriate environmental conditions. The life cycle in degradation of bioplastic is the same as a natural based materials [1].

PLA is aliphatic polyester synthesized from lactic acid in sugar and starch based crop fermentations such as corn, sugarcane, and cassava. PLA is categorized in thermoplastic which is useful for glass, film, and food packaging. For this type of bioplastic, landfill is considering as the waste disposal method employing biological enzymes, oxidation, hydrolysis, photolysis, and radiolysis in natural degradation [2]. At the end of process, carbon dioxide, water (aerobic condition), and methane (anaerobic condition) are generated as a product. Methane content in produced mixed gases called biogas from anaerobic process is useful for renewable energy source. However, in practical, PLA was highly crystalline structures that were not easily degraded under aerobic and anaerobic mesophilic conditions [3]. From previous studies, it was reported that thermophillic digestion condition could increase the biodegradation rate. However, the energy consumption of digestion process in long term operation should be carefully considered [4].

Therefore, the short time experimental period of thermal pretreatment is tempting. This could be used to enhance the biogas production and biodegradation of PLA by increasing the accessibility of micro-organisms in biodegradation process. Moreover, alkaline pretreatment involed in changing of PLA to lactic acid. Thereby, lactic acid is converted into methane. Moreover, the rising of temperature also increase the hydrolysis [5]. However, the combination effect of thermal and alkaline solution on PLA degradation is limited. In this study, an attempt is focused to investigate the effect of alkaline concentration, pretreatment temperature and reaction time on the degradation and biogas production from PLA film under mesophilic condition.

2. Materials and Methods

2.1. Raw materials
In this work, the degradation PLA films was study. It was prepared from commercial PLA pellets by blown film process. The PLA film was 80 μm thickness. The glass transition temperature (T_g) was 55-60 °C and melt temperature (T_m) was 145-160 °C. The PLA films were first cut into 3-5 mm with a grinder. Milled samples were analyzed for major chemical components, such as total solids (99.78%) and total volatile solid (99.78%).

2.2. Experimental designed conditions
The PLA pretreatment conditions were optimized by using Central Composite Design (CCD) under Response Surface Methodology (RSM) to determine the influence of several variables. That was combined with mathematical and statistical based theory wherein interactions between variable can be recognized with fewer experimental runs. The RSM also provides an optimization of pretreatment condition to offer an economic side. The CCD contained 20 experiment runs. In practical, the NaOH concentration, temperature and reaction time were used as the independent variables on biogas production, sCOD removal (%) and biodegradability (%). Within the experimental, NaOH concentration, temperature and reaction time were ranged of 0-0.5 M, 20-60 °C and 0-24 hr, respectively. The real values from CCD used in the experiment design are presented in Table 1.

Table 1. The real values used in the experimental composite design

True value		
A : NaOH conc. (M)	B : Temperature (°C)	C : Reaction time (hr)
0*	20	0*
0	30	0
0.25	45	12
0.50	60	24
0.67	70	32.2

* the value was analyzed from program is negative then assigned to zero.

2.3. Thermal-alkaline pretreatment
Thermal-alkaline pretreatment was conducted by soaking the grinded PLA films in sodium hydroxide solution at 10% (w/v) of PLA to solution. The individual variable effects of NaOH solution (0-0.67 M), temperature (20-70 °C) and reaction time (0-32.2 hr) were used according to a CCD setup. For temperature effect, the sample was incubated in the controlled temperature water batch.

2.4. Biochemical methane potential (BMP) assay
An assay called BMP was conducted. This experiment was used to determine the amount of accumulated produced biogas and sCOD removal (%) under 90 days incubation at mesophilic

temperature (30±2 °C). BMP tests were conducted in duplicate. Then, the pretreated PLA were added to the serum bottles with inoculums, obtaining a substrate: inoculum ratio was set to 60:40 on working volume and inoculum only (blank). Each bottles were added with PLA at a concentration of 2% VS based and 2.5% VS of inoculum. Fresh anaerobic sludge from UASB digester treating soft-drink was applied. The bottles were shacked with 165 rpm for 90 days at room temperature at 30±2 °C. Produced biogas was routinely monitored.

2.5. Analysis

Soluble COD (sCOD) was determined by closed reflux method according to the standard procedure. Gas measuring syringe used for determination of gas volume produced. The biodegradation (%) was calculated comparing to the theory as indicated by equation (1) and (2) [3].

$$C_nH_aO_b + \left(n - \frac{a}{4} - \frac{b}{2}\right) H_2O \rightarrow \left(\frac{n}{2} + \frac{a}{8} - \frac{b}{4}\right) CH_4 + \left(\frac{n}{2} - \frac{a}{8} + \frac{b}{4}\right) CO_2 \quad (1)$$

$$\text{Biodegradation } (\%) = \frac{\Sigma V(s) - \Sigma V(b)}{\Sigma V(max)} \times 100 \quad (2)$$

$\Sigma V(s)$ is the CO_2 and CH4 volume from the sample bottles under ambient conditions (at 30 °C, 1 atm) in milliliters. $\Sigma V(b)$ is the CO_2 and CH_4 volume from the blank under ambient conditions in milliliters. $\Sigma V(max)$ is the maximum theoretical volume of biogas (CO_2 and CH_4) evolved after complete biodegradation of the test material under ambient conditions in milliliters (calculated from equation (1)).

3. Results and discussion

3.1. Effect of NaOH pretreatment

The results clearly demonstrated that NaOH was an important parameter for PLA degradation enhancement. At the same temperature of 45 °C and 60 °C, it was found that biogas production were increased with the concentration of NaOH increase. At 0.5 M of NaOH pretreated PLA, the produced biogas was 1.3 times (164.74 ml/gVS$_{added}$) and 1.7 times (212.86 ml/gVS$_{added}$) higher than gas production from non-pretreated condition at 0 M of NaOH. Similar result was also observed for sCOD removal (%) and biodegradability (%), which were 1.3 times (51 and 67.4%) and 1.3 times (11.27 and 15.03%), respectively. Therefore, the alkaline pretreatment process can positively affect on biogas production and biodegradation. As reported in previous literature, alkaline pretreating PLA would be a beneficial for changing PLA to lactic acid by increasing rate of hydrolysis process [5]. Lactic acid can be subsequently fermented into biogas. In addition, the alkaline pretreatment of PLA can be an efficient way to increase biogas production and biodegradation because it can increase the accessibility of microorganisms.

3.2. Effect of reaction time

The increase of reaction time was also affected the biogas production, sCOD removal (%) and biodegradability (%). The maximum reaction time of 32.2 hr (0.25 M of NaOH, 45 °C) were produced the maximum biogas production, sCOD removal (%) and biodegradability (%) which were 230.21 ml/gVS$_{added}$, 74.68% and 21.62%, respectively. This result showed that reaction time was related to the increasing degradation of PLA. Similarly, Chauliac (2013) [5] found that the highest yield of lactic acid was obtain when increasing of reaction time due to the NaOH can degraded PLA more completely

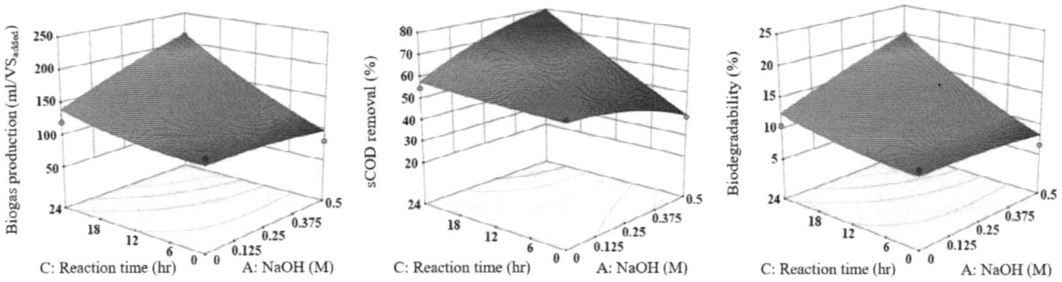

(A) on Biogas production **(B)** on sCOD removal **(C)** on Biodegradability

Figure 1. 3D response surface plot by using RSM method: Effect of NaOH concentration and reaction time at 60°C on Biogas production (A), on COD removal (B) and on Biodegradability (C)

3.3. Effect of thermal pretreatment

From the result, it was found that the temperature was less significant affect to PLA degradation. The biogas production, sCOD removal and biodegradability were slightly increased when increasing of the temperature. For example, the different temperature (20, 45, 70 °C) at NaOH concentration of 0.25 M and reaction time of 12 hr, the biogas production were 126.15, 132.42 and 147.14 ml/gVS$_{added,}$ respectively. However, the high temperature condition has been reported to enhance PLA degradation [4]. At 60 °C with NaOH solution can increase the reaction and the amount of lactic acid released from PLA. Since 60°C is not high enough to completely melt the PLA beads within the four hour, NaOH access is restricted to the amorphous layer on the bead surface [5]. The RSM analysis was presented in figure 1 and 2, that calculated for the optimum condition with providing maximum biogas production, sCOD removal (%) and biodegradability which were 0.5 M of NaOH, 60 °C and a 24 hr of reaction time.

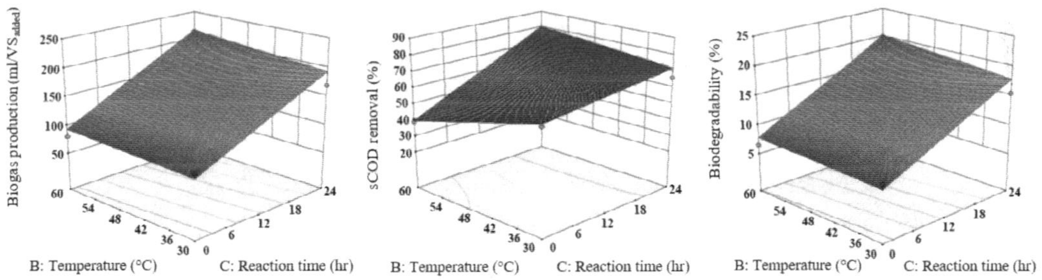

(A) on Biogas production **(B)** on sCOD removal **(C)** on Biodegradability

Figure 2. 3D response surface plot by using RSM method: Effect of temperature and reaction time at 0.5 M of NaOH on Biogas production (A), on COD removal (B) and on Biodegradability (C)

In discussion, the optimization pretreatment condition by using RSM analysis were estimated. RSM method with quadratic regression equation for describing the effects of the three independent variables was applied. The result showed that the highest biogas production was achieved at 215.47 ml/gVS$_{added}$, while sCOD remove efficiency and biodegradability of 79.67% and 20.14%, respectively, at the highest desirability of 0.93, 0.5 M of NaOH, 60 °C and a 24 hr of reaction time. From the ANOVA result of quadratic regression model for 2 responses (NaOH concentration and reaction time) demonstrated that the models were significant (P-values< 0.05). For effect of temperature shown that it did not significantly affect on biogas production (P-values > 0.05), sCOD removal, and biodegradability. Final equation of independent and dependent variables in terms of coded factors (Equation (3)-(5)):

$$\text{Biogas production (ml/gVS}_{added}\text{)} = 134.69 + 14.23A + 6.03B + 32.93C + 1.48AB + 22.15AC + 3.94BC - 3.99A^2 - 3.18B^2 + 7.20C^2 \quad (3)$$

$$\text{sCOD removal (\%)} = 60.12 + 6.27A + 0.96B + 3.21C - 4.42AB + 9.06AC + 7.65BC - 3.29A^2 \quad (4)$$

$$\text{Biodegradability (\%)} = 12.01 + 1.43A + 0.61B + 3.31C + 0.15AB + 2.23AC + 0.40BC - 0.40A^2 - 0.32B^2 + 0.72C^2 \quad (5)$$

A result from scanning electron microscopy (SEM) of non-pretreated PLA films and pretreated PLA films shows in Figure 3 and 4. It was shown that using the thermal-alkaline pretreatment, the surface of PLA films were significantly destroyed [5].

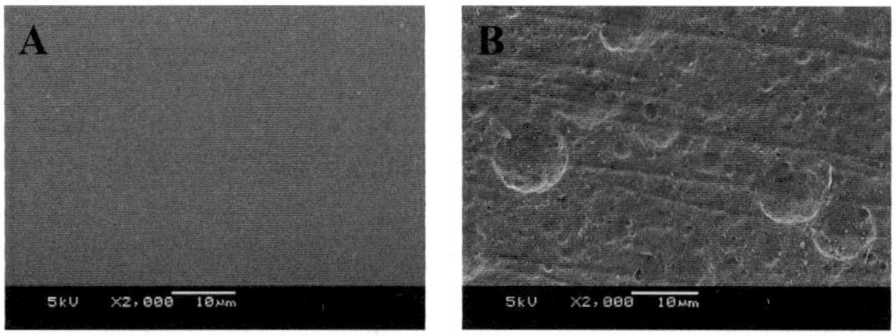

Figure 3. Surface of non-pretreated PLA films (A) and pretreated PLA films with 0.5 M NaOH at 60°C for 24 hr (B)

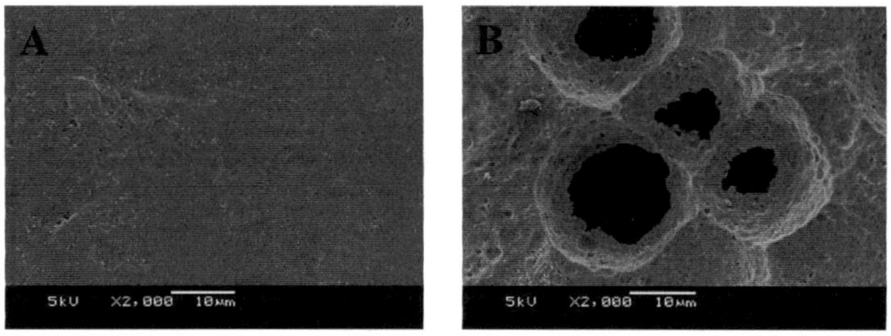

Figure 4. Surface of non-pretreated PLA film after BMP test (A) and pretreated PLA films with 0.5 M of NaOH at 60 °C for 24 hr after BMP test (B)

4. Conclusion

In this study, the RSM was applied to optimize individual variables for maximizing response. PLA is potentially used as biogas substance in anaerobic biodegradation. This research shown that reducing of hydrolytic retention time in anaerobic digestion by thermal-alkaline pretreatment. This process used 0.5 M of NaOH at 60 °C for 24 hr. The highest biogas production was 215.47 ml/gVS_{added} which was 3.7 times higher than the non-pretreated PLA films (58.28 ml/gVS_{added}) in 90 days.

References

[1] Papong S, Malakul P, Trungkavashirakun R, Wenunun P, Chom-in T, Nithitanakul M and Sarobol E 2014 *J. Cleaner Prod.* **65** 539.
[2] Nampoothiri K M, Nair N R and John R P 2010 *J. Biores. Technol.* **101** 8493
[3] Yagi H, Ninomiya F, Funabashi M and Kunioka M 2009 *Int. J. Mol. Sci.* **94** 1397
[4] Shi B and Palfery D 2012 *Int. J. Environ. Waste Manage.* **10** 297

[5] Chauliac D S 2013 *Development of a thermochemical process for hydrolysis of polylactic acid polymers to l-lactic acid and its purification using an engineered microb* (Florida)

Acknowledgement

This research was funded by the Ratchadaphiseksomphot Endowment Fund (2016), Chulalongkorn University (CU-59-002-IC) and the 90th anniversary of Chulalongkorn Fund (Ratchadaphiseksomphot), Thailand

IOP Conf. Series: Materials Science and Engineering **222** (2017) 012010 doi:10.1088/1757-899X/222/1/012010

A complete life cycle assessment of high density polyethylene plastic bottle

P Treenate[1,a] , N Limphitakphong[2,b] and O Chavalparit[1,2,c]

[1] Department of Environmental Engineering, Faculty of Engineering, Chulalongkorn University, Bangkok 10330, Thailand
[2] Research Unit of Environmental Management and Sustainable Industry, Faculty of Engineering, Chulalongkorn University, Bangkok 10330, Thailand

[a] patthamaporn.aim@gmail.com , [b] nantamoll@gmail.com , [c] orathai.c@chula.ac.th

Abstract. This study was aimed to determine environmental performances of a lubricant oil bottle made from high density polyethylene and to develop potential measures for reducing its impacts. A complete life cycle assessment was carried out to understand a whole effect on the environment from acquiring, processing, using, and disposing the product. Two scenarios of disposal phase; recycle and incineration: were examined to quantify a lesser degree on environmental impact. The results illustrated that major impacts of the two scenarios were at the same categories with the highest contributor of raw material acquisition and pre-processing. However, all impacts in case of recycling provided a lower point than that in case of incineration, except mineral extraction. Finally, feasible measures for reducing the environmental impact of high density polyethylene plastic bottle were proposed in accordance with 3Rs concept.

1. Introduction

Nowadays, plastic products are growing rapidly as the high demand for greater use. Among the variety of plastics, high density polyethylene (HDPE) was launched in the global market for using in daily life in various applications such as bottle, bag, packaging containers, drums, toys, and household appliance. To supply a greater use, a huge quantity of energy was consumed in entire life cycle of products approximately 4.27 million GJ a year and emitted a great amount of greenhouse gas (GHG) emission about 204 million ton in EU [1].

Among many techniques for estimating environmental impacts of product, life cycle assessment (LCA) defined by the International Organization for Standardization in the ISO 14040-44:2006 was considered as a valid tool to be used because it is the compilation and evaluation of the inputs, outputs, and the potential environmental impacts of a product system throughout its life cycle, which can help to understand the environmental impacts of product from the acquisition of raw materials to final disposal [2]. This methodology was widely applied in many products; for instance, textile, building, and food [3,4,5].

Despite environmental impact assessment of plastic products was interested in many researches, most of them were aimed at determining only GHG emissions in a scope of cradle to gate [6,7,8]. In this study, therefore, environmental impacts of a lubricant oil bottle made from HDPE was evaluated through life cycle assessment tool in a scope of cradle to grave, covering raw material acquisition, plastic bottle production, use, and waste disposal. Moreover, two scenarios in the end-of-life phase; recycling and incineration, based on a fact pattern were examined to emphasize the difference of environmental performances on a variety of waste management. Consequently, feasible measures will be proposed according to the findings of this study for leading to the global goal of growing green properly.

2. Methodology

LCA, a method to evaluate environmental impacts in a whole life cycle of product, was used in this study for estimating the environmental performances of lubricant oil bottle made from HDPE. Following ISO 14040-44:2006, the methodology is divided into four steps as follows:

2.1. Goal and scope definition

Firstly, goal of this study was quantifying the level of environmental performances of a lubricant oil bottle made from HDPE to understand a hotspot and a key factor in entire process. Two scenarios of waste management based on a fact pattern were established to emphasize the suitable approach for handling the lubricant oil bottle waste in Thailand. The former was recycle which was introduced by waste management company and by observing user behavior, while the latter was incineration which conformed to the governmental policies as the product type of lubricant oil bottle was declared as a hazardous waste.

Functional unit, a reference unit for collecting the inventory data, was a lubricant oil bottle with capacity of 6 liters or 350 grams, and boundary of this study covered all stages of life cycle related to the HDPE bottle production including raw material acquisition, ethylene production, HDPE pellet production, plastic bottle manufacturing, use and waste disposal as demonstrated in Fig. 1.Some text.

2.2. Inventory analysis

Life cycle inventory data of resources, material and energy consumption of lubricant oil bottle classified by phase of life cycle as raw material acquisition and pre-processing, plastic bottle production, bottle usage, and bottle waste disposal were collected from relevant sources. Firstly, raw material data obtained from olefin and HDPE plants. Secondly, plastic processing plant provided the inventory data during a bottle production phase. Thirdly, using the bottle cause no additional harm, there were neither data nor impacts for estimating. Lastly, the data of waste disposal phase was referred to the newly released ecoinvent V.3 database.

As shown in Fig. 1, the lubricant bottle with weight of 0.350 kg was produced from 0.385 kg of HDPE pellets by blow molding process. Approximately 0.1% of this amount would become wastes which were transported to external recycling plant. Furthermore, it was clearly observed that the greatest amount of energy was required for the olefin plant.

2.3. Impact assessment

The impact assessment was performed by SimaPro V.8.2 using IMPACT2002+ V.2.12 method. The results would be expressed a quantity of 15 midpoint environmental impact categories per the functional unit. In addition, the normalization was applied to analyze the contribution of each factor to the environmental performance with different scenarios of waste management.

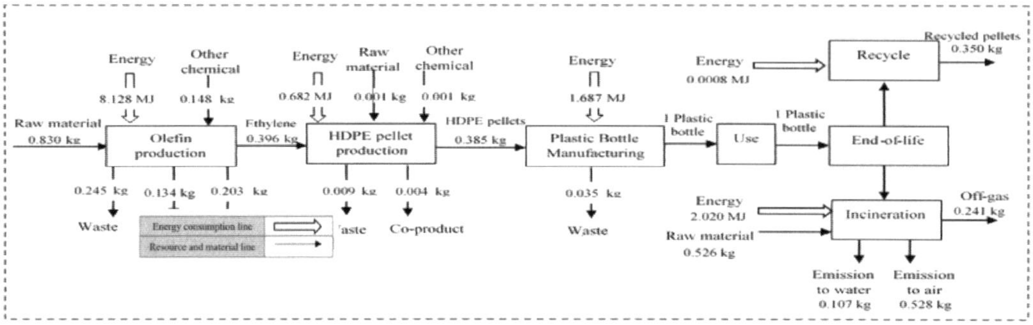

Figure 1. System boundary of a lubricant oil bottle

2.4. Interpretation

The findings of the impact assessment would be interpreted to reveal a critical point of environmental performances both in term of a major impact and a main contributor. Of what findings in this study, feasible measures for reducing the environmental impact of high density polyethylene plastic bottle would be proposed in accordance with 3Rs concept. However, the measures should be practical and consider the cost-benefit ratio in more detail.

3. Results

LCA, a method to evaluate environmental impacts in a whole life cycle of product, was used in this study for estimating the environmental performances of lubricant oil bottle made from HDPE. Following ISO 14040-44:2006, the methodology is divided into four steps as follows:

3.1. Environmental impacts

In this study, a complete cradle-to-grave LCA of lubricant oil bottle made from HDPE was presented in two scenarios which differed in the end-of-life phase; recycle and incineration. Fifteen midpoint impact categories of one HDPE bottle were presented in Fig. 2. It was found that the level of all impact categories in scenario 1 were less than that in case of disposal by incineration except mineral extraction. Therefore, it could be concluded that disposal by recycling is a better method for reducing environmental impact of plastic production.

The normalized results of two scenarios in this study was shown in Fig. 3(a). It emphasized a level of difference of environmental impacts between the two scenarios which focused only on a group of major categories including non-renewable energy, global warming, carcinogens, non-carcinogens, respiratory organics, and respiratory inorganics. The others impact categories were omitted as a cut-off rule of 95% based on the level of environmental significance [9].

Based on the majority of environmental performances, total point of six impact categories were summarized and classified the contributions among 4 contributors. Because the term of raw material acquisition and pre-processing in this study covered ethylene production and HDPE pellet production, as shown in Fig.1, this phase, therefore, was a hotspot of plastic bottle production obviously with a contribution of 80% in scenario 1 and of 60% in scenario 2 as shown in Fig. 3(b).

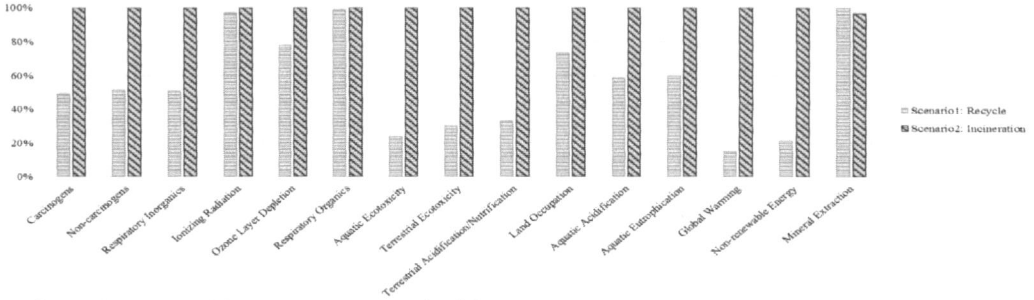

Figure 2. Environmental impact assessment in fifteen impact categories using IMPACT2002+ method

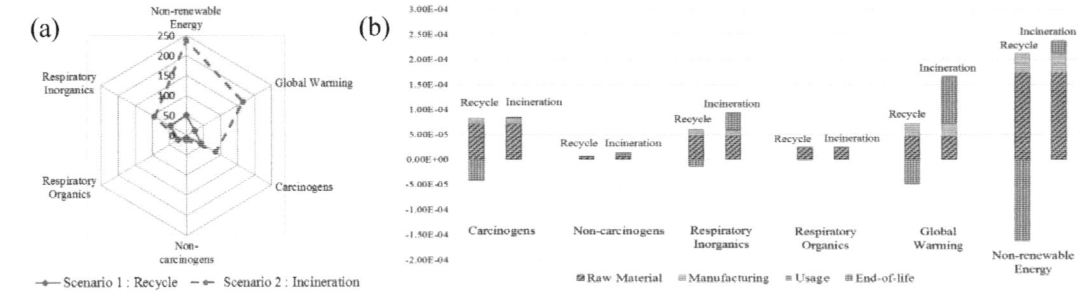

Figure 3. Normalization results: (a) Level of environmental impacts and (b) Environmental impacts contribution in different scenarios of waste management

3.2. Comparison of GHG emissions with other researches

From literature review, it was found that most researches have studied the environmental performances of plastic bottle made from polyethylene terephthalate (PET), which plastic is not a same type as conducted in this study [10,11]. However, Wrap (2010) revealed an impact of GHG emissions of milk bottle made from HDPE at 2.74 $kgCO_2eq./kg$ of HDPE pellet in case disposal by recycle and at 4.70 $kgCO_2eq./kg$ of HDPE pellet in case disposal by incineration [12]. To compare the results of such research with this study, only impact of climate change would focus only in term of $kgCO_2eq./kg$ of HDPE pellet, which differed from the functional unit as mentioned above. After recalculating, it was found that GHG emissions in case of disposal via recycle of this research (0.89 $kgCO_2eq./kg$ of HDPE pellet) was less than the point mentioned above. It might be because the different types of raw material. Lubricant oil bottle in this research used ethane as cleaner raw material, while others used a various kind of raw material such as naphtha and LPG [13]. In case of disposal by incineration, however, the result of this study displayed a higher value of impact (5.48 $kgCO_2eq./kg$ of HDPE pellet) than the point mentioned by Wrap because the product was treated as a hazardous waste whilst it was a municipal waste for the other cases.

4. Discussion

As the results of life cycle assessment, non-renewable energy was a main issue due to the use of natural gas as raw material for producing ethylene to use in HDPE production consequentially and that made the phase of raw material acquisition and pre-processing as a hotspot. To reduce environmental impacts of lubricant oil bottle, therefore, feasible measures should be focused on the effect on non-renewable energy and the impact of raw material acquisition deeply. The 3Rs concept, the basis of sustainable waste management was applied and proposed in this study as follows:

- Non-renewable energy reduction. Currently, energy security in Thailand is declined because it ran out of non-renewable energy and relied heavily on imported energy. A promotion of alternative energy development such as solar, wind, hydro power, biomass, biogas, municipal solid waste, and biofuels (ethanol and biodiesel), therefore, is crucial [14].
- Energy conservation. Improving energy efficiency in the process at plant through any facilities and/or equipment does not only help saving an amount of energy but also help cutting off some energy cost and decreasing an impact of global warming as well [6]. Some instances of energy conservation measure including controlling a percentage of excess oxygen in furnace and changing fan blade should be promoted and implemented [15].
- Recycle of raw material. Although, bioplastic is another option for reducing both raw material origin and GHG emission, however, it is costly and might affect food security [16]. Plastic product and its scrap should be recycled as raw material for producing plastic bottle. It is beneficial not only in reducing raw material origin but also reducing fuel consumption and waste generated [17].

- Waste management. Another option for minimizing consumption of natural resources was sustainable industrial waste management toward zero landfill waste. As shown in Fig.1, the biggest amount of waste generated from olefin plant of which 30% of raw material would become waste such as activated carbon, coke, and oligomer. The main options for waste management were selling to recycling plant, reusing and recycling, and industrial waste landfilling [18].

5. Conclusion

This research was aimed at estimating the environmental impacts in the entire life cycle of high density polyethylene lubricant oil bottle and developing feasible measures for reducing its impacts. Two different scenarios in the end-of-life phase; recycle and incineration: were examined to identify a suitable alternative of environmental performances for plastic bottle production in Thailand. The results revealed total environmental impact from entire life cycle of lubricant oil bottle made from HDPE provided lesser point in recycle scenario than disposing by incineration. A majority in both scenarios was the same category even differed in sequence including non-renewable energy, global warming, carcinogens, non-carcinogens, respiratory organics, and respiratory inorganics. Based on the majority, a hotspot of plastic bottle production obviously occurred in the phase of raw material acquisition and pre-processing with the main impact of non-renewable energy. Therefore, the recommendations for reducing environmental performances of plastic bottle product would be mainly focused on reducing non-renewable energy consumption and minimizing a number of origin materials by using alternative energy, applying energy conservation measure, recycling plastic product and its scrap as raw material and managing industrial waste properly which will contribute to the sustainable development product.

References

[1] Pilz H, Brandt B and Fehringer R 2010 *The impact of plastics on life cycle energy consumption and greenhouse gas emissions in Europe* (Vienna: Denkstatt)
[2] Siracusa V, Ingrao C, Giudice A L, Mbohwa C and Rosa M D 2014 *Food Res. Int.* **62** 151-61
[3] Yaseneva P et al. 2014 *Chem. Eng. J.* **248** 230-41
[4] Kylili A, Ilic M and Fokaides P A 2017 *Resour., Conservat. and Recycl.* **116** 169-177
[5] Baldini C, Gardoni D and Guarino M 2017 *J. Cleaner Prod.* **140** 421-435
[6] Imardoon K, Chavalparit O and Varabuntoonvit V 2012 *Proc. Int. Conf. on Environmental Science, Engineering and Management* (Bangkok: Environmental Engineering Association of Thailand)
[7] Tsiropoulos I, Faaij A P, Lundquist L, Schenker U, Briois J F and Patel M K 2015 *J. Cleaner. Prod.* **90** 114-27
[8] Zhao Z, Liu Y, Wang F, Lic X, Deng S, Xua J, Weid W and Wanga F 2016 *J. Cleaner. Prod.*
[9] European Commission 2016 Guidance for the implementation of the EU Product Environmental Footprint (PEF) during the Environmental Footprint (EF) pilot phase version 5.2 (Brussels: European Commission)
[10] Garfí M, Cadena E, Sanchez-Ramos D and Ferrer I 2016 *J. Cleaner. Prod.* **137** 997-1003
[11] Kanga D H, Aurasa R and Singh J, 2017 *Resour., Conservat. and Recycl.* **116** 45-52
[12] Fry J M, Hartlin B, Wallén E and Aumônier S 2010 *Life cycle assessment of example packaging systems for milk* (Banbury: Waste & Resources Action Programme)
[13] Chuapet W, Limphitakphong N, Tantisattayakul T, Kanchanapiya P and Chavalparit O 2016 *Proc. Int. Conf. on Industrial Engineering and Applications* vol 68 (Hongkong: EDP Sciences)
[14] Ministry of Energy 2013 *Energy in Thailand facts & figures 2013* (Bangkok: Department of Alternative Energy Development and Efficiency, Ministry of Energy)
[15] Ministry of Energy 2007 *A Study on Energy Efficiency Index in Petrochemical Industry* (Bangkok: Department of Alternative Energy Development and Efficiency, Ministry of

Energy)
[16] Yu J and Chen L X L 2008 *Environ. Sci. Technol.* **42** 6961–66
[17] Achilias D S, Roupakias C, Megalokonomos P, Lappas A A and Antonakou E V 2007 *J. Hazard. Mater.* **149** 536–42
[18] Usapein P and Chavalparit O 2014 *Mater. Cycles Waste Manag.* **16** 373-83

Acknowledgement
This research was funded by the Ratchadapisek Sompoch Endowment Fund (2014), Chulalongkorn University (CU-57-062-CC). The author also would like to gratefully thank all the factories for providing the collaboration in this research.

IOP Conf. Series: Materials Science and Engineering **222** (2017) 012011 doi:10.1088/1757-899X/222/1/012011

Special features of design and calculation for structures made of anisotropic fiberglass

S L Shambina[1,3] and G A Virchenko[2]

[1] Department of Architecture and Civil Engineering, Academy of Engineering, Peoples' Friendship University of Russia, 6, Miklukho-Maklaya Str., Moscow, Russia, 117198
[2]National Technical University of Ukraine "Kiev Polytechnic Institute", Pobedy ave., 37, Kyiv, Ukraine, 03056

Email: [3]shambina_sl@mail.ru

Abstract. In recent years composite materials find wide application in various fields of engineering, because they have a number of advantages over other structural materials. A variety of composites' physical and mechanical properties (especially anisotropy) requires an improvement of existing calculation methods and creation of new ones for structural elements made of these materials. This is an important task which will contribute to their wider use. In this paper some famous criteria of anisotropic materials are examined, and their advantages and disadvantages are discussed. The authors of the paper suggest new variants of strength criteria for anisotropic materials. These new criteria are based on new mechanical characteristics which are more convenient for experimental obtaining. Also new criteria use separate form of writing for each quadrant of the stress plain.

1. Introduction

Composite materials, primarily fiberglass, have their special place among all the materials used today in construction. They have the most attractive prospect because of their physical and mechanical characteristics. For example strength of fiberglass (glass-reinforced plastics or GRP) is comparable to strength of steel, their thermal conductivity is equal to the one of wood, and its ease is similar to the ease of PVC. Composite materials combine the best qualities of traditional materials but without their disadvantages. A large class of structures which have been usually made of metal and concrete, aluminium and PVC-plastic (polyvinylchloride) can be successfully and efficiently transferred to the use of fiberglass.

2. Basic properties of fiberglass plastics

Fiberglass is a composite material consisting of two main components which are polymer (binder) and glass fibers (filler). Various combinations of these components allow adjusting the properties of glass-reinforced plastics in a wide range. These materials have a lot of excellent properties such as: high specific strength, relatively low weight, low thermal conductivity, high chemical stability and damping capacity, the ability to withstand high temperatures for a short time, high workability, high fatigue strength and others [1]. Fiberglass is not subjected to corrosion and rot, do not require painting; some of them pass up to 90% of light and 70% of ultraviolet rays. Strength of some fiberglass exceeds strength of the best structural steels and aluminium alloys, and it is much more than the one concrete and reinforced concrete.

3. Application of fiberglass in construction

Fiberglass gives the opportunity to create structures of almost any shape. It is efficient to use fiberglass plastics in different types of building structures: covering plates, shells, domes in translucent constructions, in lightweight long-span structures, in special radio transparent structures, as well as in the construction on subsiding soil or in seismic areas.

In spatial elements the relatively low modulus of elasticity of fiberglass plastics is compensated by the increased stiffness and stability of dimensional systems as well as by the use of rational cross sections. Spans of spatial constructions made of fiberglass can reach 50 meters with a small material consumption and low weight of structures.

It is very perspective to use fiberglass for highly loaded elements and structures, and for those which are operating under impact shock loads.

Also fiberglass is used for manufacturing pipes. Main advantages of such pipes in comparison with metal ones are their high strength, resistance to corrosion from action of gases, salt solutions, acids and alkalis. Such pipes are lightweight (4 - 5 times less than weight of similar steel pipes), have low thermal conductivity. They are used in plumbing, in oil industry, as a lining shaft, for the manufacture of rod systems for building structures.

An important advantage of fiberglass plastics is the ability to use them for effective multi-layered structures such as sandwich plates and shells. In such structures the outer layers are made of fiberglass plastics, and inner layers are made of foamed polystyrene or sotoplast or goffered aggregates. The combination of layers having different properties makes it possible to provide a reliable structure for work in adverse environmental conditions. These structures have relatively low weight and good heat insulation, electrical insulation and sound insulation properties. Multi-layer constructions provide structure's mass reduction by 12 - 40% with set limits on the strength and stiffness. At the same load the bearing capacity of panels with composites is 10 - 65% lower than that of panels with metal skins. The use of such panels makes it possible to reduce the load on the supporting structures and therefore decrease the weight of trusses, pillars and foundations, as well as reduce the cost of transportation, installation, operation, which provides additional savings.

The most widely varied opportunities of fiberglass plastics are disclosed in public buildings and facilities, allowing the creation of original designs, as fiberglass combines both structural and protecting functions, and sometimes lighting function. Dimensional fiberglass structures are used for erection of pavilions coverings, sports facilities, indoor markets and others.

In the manufacture of the composites may be used various techniques (extrusion, contact molding, extrusion, casting, winding).

4. Anisotropy of the physical and mechanical properties of fiberglass

It is a very important feature of fiberglass plastics made on the basis of glass fabrics or directed glass strands. Anisotropy is a dependence of the material's characteristics of the angle between the directions of the reinforcement and operating stresses. Elastic and strength properties of fiberglass (including their long-term strength, creep and others) have the feature of anisotropy. Anisotropy of strength means that the fiberglass tensile strength, compressive or shear strength depends on the orientation of the glass fibers in the sample. In other words, anisotropy means the difference of properties in different directions. Anisotropy of fiberglass plastics can be a technological, i.e. appearing in the process of manufacturing of elements, and a constructive i.e. defined during designing of details.

A study of anisotropy of fiberglass's elastic and strength properties has a great theoretical and practical interest [2]. Using the correct choice of the anisotropy's character in fiberglass it is possible to create lightweight structures of equal strength. This is a very important advantage of GRP compared to many other construction materials. It is possible to use the material in a structure in an optimal way by changing the order of layers and direction for stacking threads in the layer. The strongest materials can

IOP Conf. Series: Materials Science and Engineering **222** (2017) 012011 doi:10.1088/1757-899X/222/1/012011

be obtained if they are reinforced by straight threads that have been laid with pretension. Depending on the location and type of fiber, there are various types of reinforced plastics. The most advantageous in terms of improving the strength is location of reinforcement in the direction of the greatest tensile forces. However, in most cases, the best option of reinforcement type can be find only by calculations. When studying the anisotropy it is convenient to use methods of the tensor calculus.

The topical is the question of the strength calculation for GRP structural elements under *complex stress state*, taking into account the materials' features of mechanical properties. The real physical model of a composite's work in a structure is complicated, it is necessary to choose a proper approximating model to simplify practical calculations of strength and stability. Currently, a *phenomenological approach* is widespread in the mechanics of composites, when the composite material is considered as a continuous and homogeneous anisotropic medium with averaged macro stresses and strains.

5. Strength criteria for anisotropic materials

For practical calculations of elements made of anisotropic materials it is necessary to use any criterion of strength for anisotropic materials [3], [4]. The basic characteristics of the material included in the strength criteria, should be obtained from experiments in simple forms of strain-stress condition.

There are quite a lot of phenomenological strength criteria for anisotropic materials, which have different graphic interpretation in the stresses space. Common approach to the description of a strength surface does not exist, so the choice of this or that criterion is always a relatively subjective task. The particular form of expression taken as a strength criterion for a material can be verified only experimentally. The best known criteria are the following phenomenological strength criteria for anisotropic materials: ones named by Mieses - Hill, Zakharov - Malmeyster, Goldenblat – Kopnov, etc.

Despite the large number of strength criteria for anisotropic materials, their application in practice of design is rather limited for several reasons:

1) None of the known strength criteria is universal, i.e. they all have certain limits of applicability. The most universal criterion from this point of view is Goldenblat - Kopnov criterion.
2) Analytic expressions of some criteria of strength are rather cumbersome and contain many basic material strength characteristics (ultimate tensile strength, compression and shear ones).
3) It is technically difficult to determine experimentally some parameters of the ultimate strength (i.e. ultimate shear strength in directions at angle 45° with respect to the axes of anisotropy).

The task of creating of strength criteria for anisotropic materials (such as reinforced plastics) has been solved in two ways: by generalization of the well-known strength criteria of isotropic materials and by development of essentially new criteria.

V. Kopnov and S. Shambina suggested several new variants of known strength criteria without the mentioned above drawbacks. It was suggested to use new basic strength constants of materials. Also the authors proposed new approach to some known strength criteria; it was suggested to write these criteria separately for each quadrant of stress plane.

Let us consider in more details how the new modifications Mieses- Hill criterion for anisotropic materials were obtained. This criterion has been get on the basis of the plasticity criterion proposed by Mieses for metals. For the case of plane stress state Mieses- Hill criterion for anisotropic materials is the following:

$$\frac{\sigma_{11}^2}{\sigma_{T1}^2}+\frac{\sigma_{22}^2}{\sigma_{T2}^2}+\frac{\sigma_{12}^2}{\tau_{T12}^2}-\left(\frac{1}{\sigma_{T1}^2}+\frac{1}{\sigma_{T2}^2}-\frac{1}{\sigma_{T3}^2}\right)\cdot\sigma_{11}\cdot\sigma_{22}=1. \quad (1)$$

where σ_{T1}, σ_{T2} - the limits of material yield stress in the material's first and second main directions; τ_{T12} - the limit of material yield stress in pure shear; σ_{11}, σ_{22}, σ_{12} - the values of the current stresses.

In [5] it was suggested to express the coefficients of Mieses-Hill criterion through the limits of tensile (compressive) strength and the limit of shear on the site situated at the angle of 45 ° with the main direction of the material ($\tau_{в45}$) and the limit of shear on the sites coinciding with the main directions of the material ($\tau_{в12}$).

For the case of plane stress state this criterion has the form:

$$\frac{\sigma_{11}^2}{\sigma_{в1}^2}+\frac{\sigma_{22}^2}{\sigma_{в2}^2}+\left(\frac{1}{\sigma_{в1}^2}+\frac{1}{\sigma_{в2}^2}-\frac{1}{\tau_{в45}^2}\right)\cdot\sigma_{11}\cdot\sigma_{22}+\frac{\sigma_{12}^2}{\tau_{в12}^2}=1. \quad (2)$$

The possibility of using Mieses plasticity criterion as a strength criterion for anisotropic materials was studied in [5]. For materials with strong anisotropy this criterion is not consistent with the results of experiments, whereas this criterion satisfactorily approximates the results of experiments for materials with wear anisotropy. Obviously, this criterion should only be used to anisotropic materials having the same tensile and compressive strength.

Mieses criterion (Eq. 2) was first recorded in a tensor form in [4]:

$$П_{ikmn}\cdot\sigma_{ik}\cdot\sigma_{mn}=1. \quad (3)$$

In the main coordinate system in the case of plane stress state the criterion (Eq. 3) contains 4 tensor components of strength:

$$П_{1111}\cdot\sigma_{11}^2+П_{2222}\cdot\sigma_{22}^2+2\cdot П_{1122}\cdot\sigma_{11}\cdot\sigma_{22}+4\cdot П_{1212}\cdot\sigma_{12}^2=1. \quad (4)$$

To express the components $П_{ikmn}$ through elementary limit states (tension, compression and shear), two cases of uniaxial tension were considered: 1) $\sigma_{11}=\sigma_{в1}$; $\sigma_{22}=0$; $\sigma_{12}=0$; 2) $\sigma_{22}=\sigma_{в2}$; $\sigma_{11}=0$; $\sigma_{12}=0$. Substituting them in Eq. 4 we can get the following formulas for $П_{1111}$ and $П_{2222}$:

$$П_{1111}=\frac{1}{\sigma_{в1}^2};\quad П_{2222}=\frac{1}{\sigma_{в2}^2}. \quad (5)$$

After considering the shear state ($\sigma_{11}=0$; $\sigma_{22}=0$; $\sigma_{12}=\tau_{в12}$) and substituting these relations into Eq. 4, we can obtain a formula for determination of component $П_{1212}$:

$$П_{1212}=\frac{1}{4\cdot\tau_{в12}^2}. \quad (6)$$

To obtain expressions for the component $П_{1122}$ it is necessary to consider a state of pure shear ($\sigma_{11}=\tau_{в45}$; $\sigma_{22}=-\tau_{в45}$; $\sigma_{12}=0$) and substitute these relations to Eq. 4:

$$П_{1122}=\frac{1}{2}\cdot\left(\frac{1}{\sigma_{в1}^2}+\frac{1}{\sigma_{в2}^2}-\frac{1}{\tau_{в45}^2}\right). \quad (7)$$

It should be kept in mind that even small changes in value of $П_{1122}$ can significantly affect the type of strength line. Since there is not enough experimental data on the composite strength under complex stress state even for well-known composites, so it is difficult to find the value of component $П_{1122}$ that could be reliable to use in calculations of structures made of these materials. Some ways to resolve this problem will be discussed below.

6. New variants of strength criteria for anisotropic materials

In the terms of strength tensor components Eq. 5, Eq. 6, Eq. 7 there is no difference between the tensile and compressive strength. Therefore this criterion in such form can be used only for anisotropic materials with the same meanings of tensile and compressive strength in each direction ($\sigma_{вi}^{+}=\sigma_{в_i}^{-}=\sigma_{вi}$) and with equal two shear strength on areas located at an angle of 45 ° to the main direction of the material ($\tau_{в45}^{+}=\tau_{в45}^{-}=\tau_{в45}$). There are not so many of such materials available.

V. Kopnov and S. Shambina [6] suggested extending Mieses-Hill criterion for anisotropic materials with varying tensile and compressive strength. It can be achieved through using of this criterion separately for each quadrant of the plane stresses.

Also it is possible to use new material's strength characteristics instead of $\tau_{в45}$ [7]. For example, the following characteristics can be used: limits of strength under uniform of biaxial stretching

$\sigma^{++}{}_{в12}$ and uniform biaxial compression $\sigma^{--}{}_{в12}$ (with stresses ratio $\sigma_{11}/\sigma_{22}=1$) or limits of strength under biaxial stretching $\tilde{\sigma}^{++}{}_{в12}$ or biaxial compression $\tilde{\sigma}^{--}{}_{в12}$ (with a stress ratio $\sigma_{11}/\sigma_{22}=1:2$). This criterion was written separately for each quadrant of the plane stresses with new strength characteristic in the following form:

1) for quadrant $\sigma_{11}>0$; $\sigma_{22}>0$ (biaxial stretching):

$$\left(\frac{\sigma_{11}}{\sigma^{+}{}_{в1}}\right)^2+\left(\frac{\sigma_{22}}{\sigma^{+}{}_{в2}}\right)^2+\frac{1}{2}\cdot\left[\left(\frac{1}{\tilde{\sigma}^{++}{}_{в12}}\right)^2-\left(\frac{2}{\sigma^{+}{}_{в1}}\right)^2-\left(\frac{1}{\sigma^{+}{}_{в2}}\right)^2\right]\cdot\sigma_{11}\cdot\sigma_{22}=1; \tag{8}$$

2) for quadrant $\sigma_{11}<0;\sigma_{22}<0$ (biaxial compression):

$$\left(\frac{\sigma_{11}}{\sigma^{-}{}_{в1}}\right)^2+\left(\frac{\sigma_{22}}{\sigma^{-}{}_{в2}}\right)^2+\frac{1}{2}\cdot\left[\left(\frac{1}{\tilde{\sigma}^{--}{}_{в12}}\right)^2-\left(\frac{2}{\sigma^{-}{}_{в1}}\right)^2-\left(\frac{1}{\sigma^{-}{}_{в2}}\right)^2\right]\cdot\sigma_{11}\cdot\sigma_{22}=1; \tag{9}$$

3) for quadrant $\sigma_{11}>0;\sigma_{22}<0$:

$$\left(\frac{\sigma_{11}}{\sigma^{+}{}_{в1}}\right)^2+\left(\frac{\sigma_{22}}{\sigma^{-}{}_{в2}}\right)^2+\frac{1}{2}\cdot\left[\left(\frac{1}{\sigma^{+}{}_{в1}}\right)^2+\left(\frac{1}{\sigma^{-}{}_{в2}}\right)^2-\left(\frac{1}{\tau^{+}{}_{в45}}\right)^2\right]\cdot\sigma_{11}\cdot\sigma_{22}=1; \tag{10}$$

4) for quadrant $\sigma_{11}<0;\sigma_{22}>0$:

$$\left(\frac{\sigma_{11}}{\sigma^{-}{}_{в1}}\right)^2+\left(\frac{\sigma_{22}}{\sigma^{-}{}_{в2}}\right)^2+\frac{1}{2}\cdot\left[\left(\frac{1}{\sigma^{-}{}_{в1}}\right)^2+\left(\frac{1}{\sigma^{+}{}_{в2}}\right)^2-\left(\frac{1}{\tau^{-}{}_{в45}}\right)^2\right]\cdot\sigma_{11}\cdot\sigma_{22}=1. \tag{11}$$

7. Conclusions

Knowledge of mechanical characteristics for anisotropic materials and methods of calculation for structures made of these materials gives to the engineers the perfect opportunity to create reliable constructions. In this article some new variants of a well-known strength criterion were suggested. These new criteria are based on using of new strength characteristics have been obtained with the help of writing the criterion separately for each of the four quadrants. This new criterion gives the opportunity to obtain proper results for calculations of structures made of a wider range of anisotropic materials.

References

[1] Vasiliev V V 1989 *Mechanics of structures made of anisotropic materials.* – Moscow – 272 p.
[2] Noton B 1978 *Application of composite materials to technics/* In: Composite materials. Vol.3 – Moscow.: Mashinostroyeniye. – Pp. 260 - 307.
[3] Mises R 1928 *Mechanik der plastischen Formanderung von Kristallen.* ZAMM. B.8. H. 3.
[4] Hill R 1956 Mathemtical theory of plasticity. –GITTL, -407 p.
[5] Skudra A M, Bulavs F Y 1971 *Influence of tangential stresses sign on the shear strength of reinforced plastics. // Mechanics of polymers.* No. 27. Pp. 262 – 268.
[6] Kopnov V A, Shambina S L 1992 *Different representations of tensor strength components for anisotropic materials.* //Applied mechanics. – Moscow. Pp. 76 – 80.
[7] Shambina S L 2009 *Experimental checking of some modifications of Mieses-Hill criterion for anisotropic materials.* //Structural Mechanics of Engineering constructions and buildings. - Moscow. – Pp. 55 – 60.

Acknowledgement

This paper was financially supported by the Ministry of Education and Science of the Russian Federation on the program to improve the competitiveness of Peoples' Friendship University of Russia (RUDN University) among the world's leading research and educational centres in the *2016 – 2020.*

IOP Conf. Series: Materials Science and Engineering **222** (2017) 012012 doi:10.1088/1757-899X/222/1/012012

The spatial-temporal instability behavior of an electrified viscoelastic liquid sheet

Kun Jia[1,2], **Dejun Sun**[1]

[1]Department of Modern Mechanics, University of Science and Technology of China, Hefei, 230027, China

Dsun@ustc.edu.cn

Abstract. The spatial-temporal instability of an electrified viscoelastic liquid sheet injected into air in a vertical electric field is analysed, in which the complete electrical force is considered. A new mode induced by the electric field (referred to as electrical mode) is found in the long wave region. The results show that the absolute-convective instability (AI-CI) of the electrical mode is only affected by the electric field, while that of the hydrodynamic mode is significantly influenced only when the electric field exceeds a threshold value. In most cases, the electrical mode is the dominant mode unless the Euler number is very small or large. The dominant mode first transforms from the hydrodynamic mode to the electrical mode in a weak electric field, when the flow is convectively unstable. However, it would change back in absolute instability region with a strong electric field. In addition, the competition between the two modes is also influenced by the Weber number and Reynolds number.

1. Introduction

In many problems of instability, the amplitude of a disturbance develops in both time and space, which requires to use spatial-temporal instability analysis. If the amplitude of an initially localized disturbance envelope eventually decays at any fixed location while it grows when observed in a coordinate system moving with its group speed, the system is convectively unstable. Otherwise, the system is absolutely unstable [1]. A methodology for distinguishing the convective and absolute nature of the instability was applied to two-dimensional hydrodynamic instabilities by Huerre and Monkewitz [2, 3]. The AI-CI characteristic of the flow is usually determined by detecting saddle points which satisfies the Briggs-Bers collision criterion [4, 5]. If the growth rate at the saddle point, i.e. the absolute growth rate $\omega_{0i} < 0$, the flow is convectively unstable, contrarily, the flow is absolutely unstable. Recently, more and more physically consistent analysis in terms of spatial-temporal instability has been extensive used in liquid sheet. [6,7]

Recently, the study of electrified viscoelastic liquid flows has received more and more attention. El-Saied and Syam [8] investigated the electrohydrodynamic instability. The results demonstrated that increasing the Mach number from subsonic to transonic leads to the increase in the maximum growth rate and dominant wave number and the dominant wave number to increase and the electric field further accelerate their growth. Yang et al. [9] utilized a simplified linear stability model to analyse the instability characteristics of the electrified viscoelastic liquid cylindrical jet. In their work, the linear viscoelastic constitutive relation is used to describe the viscoelasticity of dilute polymer solutions under small or moderate deformation. Recently, Ruo et al. [10] performed a three-dimensional instability analysis of an electrified non-Newtonian liquid jet subjected to unrelaxed axial tension. Liu el at. [11] Studied the atomization mechanism of a charged viscoelastic liquid sheet in a stationary gas.

The effects of dimensionless parameters on the instability of the flow is studied. Lauricella [12]. investigated the effects of dissipative air drag on the dynamics of electrified jet in the electrospinning process, utilizing a Brownian noise model. They examined the role of air drag force in the electrospinning process, and provided prospective beneficial implications for improving forthcoming electrospinning experiments. Tong el at. [13] performed an investigation about the viscoelastic liquid sheet subjected to noticeable unrelaxed axial tension.

In this work, we study the spatial-temporal instability of an electrified viscoelastic liquid sheet under the complete electrical force, which is more close to practical conditions.

2. Equations

The schematic of the problem analysed is shown in figure 1.

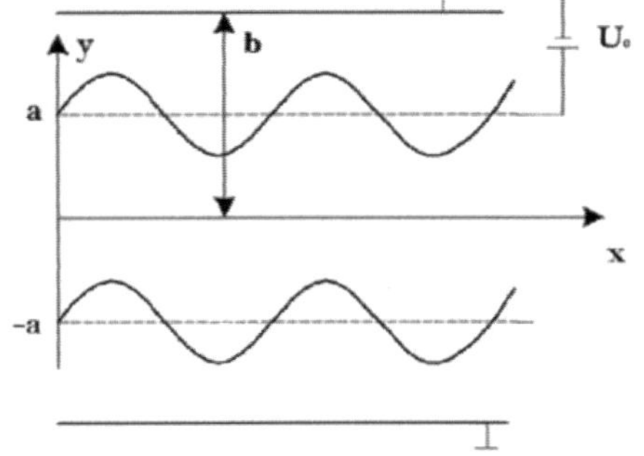

Figure 1. Sketch of the electrified viscoelastic liquid sheet. The liquid sheet has a density of ρ , a thickness of 2a and its surface tension is τ . The liquid flow moves with a horizontal velocity U. Two flat electrodes are positioned on the top and bottom surface of the liquid sheet, and the distance between them is 2b. A voltage U_0 is imposed on the sheet surface.

An electrical potential function ϕ is introduced, which satisfies the Laplace equation $\Delta\phi=0$.

The equations governing the motion of liquid in the sheet are the conservation laws of mass and momentum, i.e.

$$\nabla \cdot \mathbf{v} = 0 \tag{1}$$

$$\left(\frac{\partial}{\partial t} + U\frac{\partial}{\partial x}\right)\mathbf{v} = -\frac{1}{\rho}\nabla \cdot \boldsymbol{\pi} \tag{2}$$

Where $\boldsymbol{\pi} = p\boldsymbol{\delta} + \boldsymbol{\tau}$, $\mathbf{v}$ is the liquid velocity vector, $\boldsymbol{\pi}$ is the total stress tensor, p is the pressure, and $\boldsymbol{\tau}$ is the extra stress tensor.

The viscoelastic property of the liquid is described by the Oldroyd-B constitutive equation, i.e.

$$\boldsymbol{\tau} + \lambda_1\left(\frac{\partial}{\partial t} + U\frac{\partial}{\partial x}\right)\boldsymbol{\tau} = -\eta_0\left(\dot{\boldsymbol{\gamma}} + \lambda_2\left(\frac{\partial}{\partial t} + U\frac{\partial}{\partial x}\right)\dot{\boldsymbol{\gamma}}\right) \tag{3}$$

where $\boldsymbol{\gamma} = \nabla\mathbf{v} + (\nabla\mathbf{v})^T$ is the rate of strain tensor, and η_0 is the zero shear viscosity, λ_1 is the stress relaxation time, and λ_2 is the deformation retardation time.

The liquid sheet is perturbed by a sinuous mode infinitesimal disturbance. To linearize the equations, the normal mode decomposition is substituted. Finally, we get the dimensionless dispersion relation:

$$\frac{4\lambda K^2 We\left(1+i\mathrm{MRe}E_1(K-\Omega)\right)^2}{\mathrm{Re}^2\left(1+i\mathrm{Re}E_1(K-\Omega)\right)^2}tanh(\lambda)$$

$$-\frac{We\left(\lambda^2+K^2\right)^2\left(1+i\mathrm{MRe}E_1(K-\Omega)\right)^2}{K\mathrm{Re}^2\left(1+i\mathrm{Re}E_1(K-\Omega)\right)^2}tanh(K)+\frac{KWeE_u}{D^2}\coth(DK)-K^2=0 \tag{4}$$

set The dimensionless wave number $K = ka$, the dimensionless frequency $\Omega = (a/U)\omega$, the dimensionless parameters involved include the Reynolds number $\mathrm{Re} = \rho U a/\eta_0$, the Weber number $We = \rho U^2 a/\sigma$, the Euler number $Eu = \varepsilon_0 U_0^2/\rho U^2 a^2$, where ε_0 is vacuum dielectric constant; the elasticity number $El = \lambda_1\eta_0/\rho a^2$, the ratio of the distance between horizontal electrodes and the liquid sheet-to-sheet thickness $\mathrm{D} = (b-a)/a$, the ratio of the deformation retardation time to the stress relaxation time $M = \lambda_2/\lambda_1$. In the calculation, El = 1, D = 40, and M = 0.1. Yang et al. [17] obtained a similar dispersion relation, in which the electric field was approximated by neglecting the horizontal electrical force. However, in our dispersion relation, the complete electrical force is particularly considered.

3. Results and Discussions

A linear stability analysis is applied firstly. In figure 2, the increase of the electric field strength enhances the disturbance growth rate, and meanwhile enlarges the corresponding instability domain of wave number. In figure 3, the increase of the Reynolds number enhances the instability of the non-Newtonian fluid.

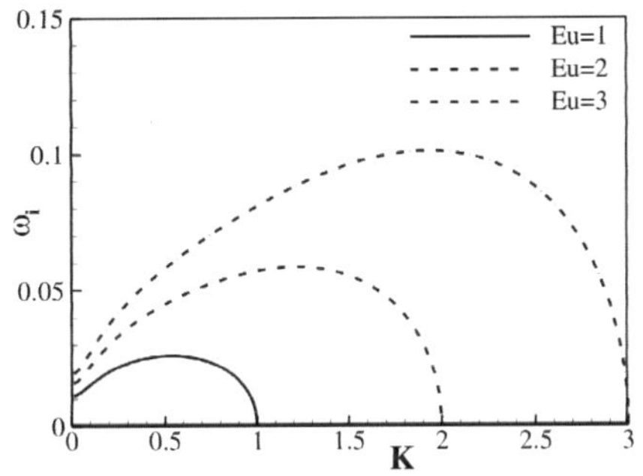

Figure 2. The effect of electric field on the disturbance growth rate of the viscoelastic fluid, where Re=1000, We=400.

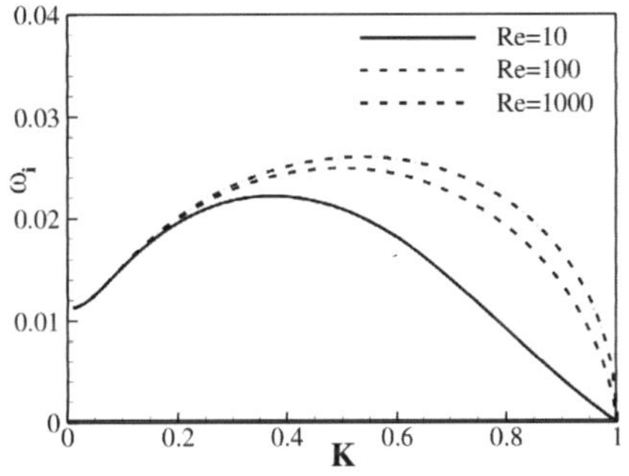

Figure 3. The effect of the Reynolds number on the disturbance growth rate of the viscoelastic fluid, where We=400, Eu=1.

In figure 4, the increase of the Weber number enhances the disturbance growth rate, and enlarges the corresponding instability domain of wave number as well.

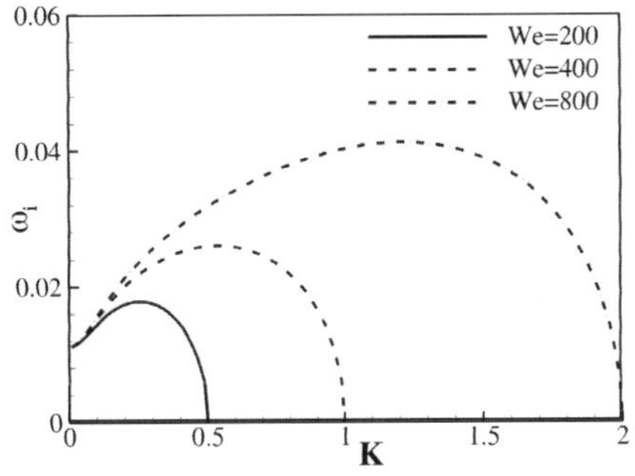

Figure 4. The effect of the Weber number on the disturbance growth rate of the viscoelastic fluid, where Re=1000, Eu=1.

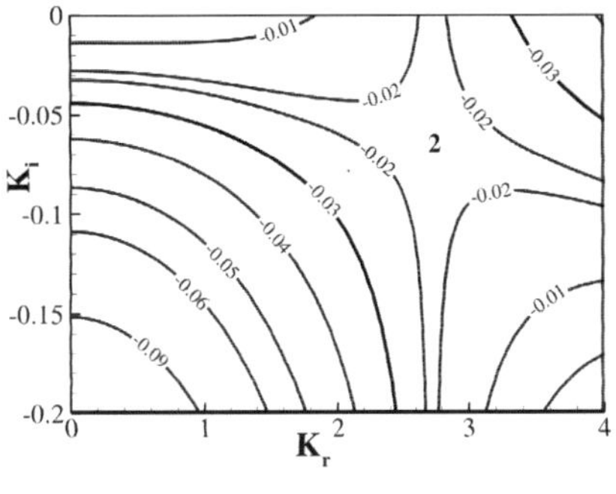

Figure 5. Contours of ω_r = constant in the complex k plane, where We = 6, Re = 80, Eu = 0.

The maps of the saddle point on the complex wave number plane for We = 6, Re = 80, Eu = 100 are shown in figure 6. A saddle point marked 1 is located in the long wave region, and the other saddle point marked 2 is located in a shorter wave region. In figure 5, when the electric field is absent, the long wave mode disappears, indicating that it is induced by the electric field, so we call it as the electrical mode.

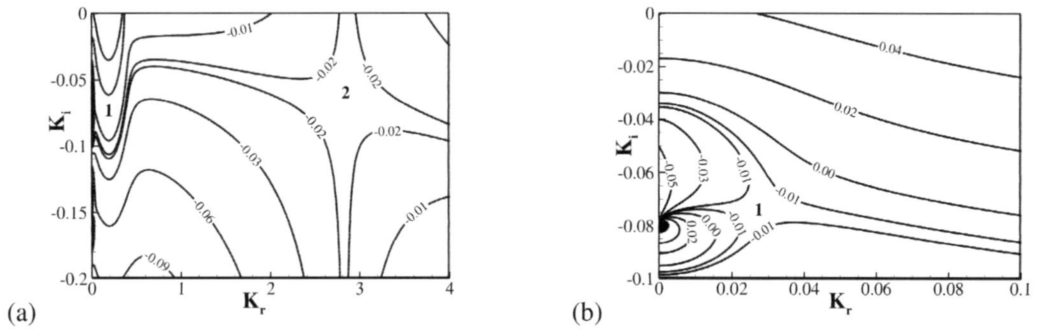

Figure 6. Maps of saddle points (a) Contours of ω_r = constant in the complex k plane, where We = 6, Re = 80, Eu = 100. (b) The amplification of small wave number region of (a).

The absolute growth rates of the two modes' the saddle points with increased electric field is shown in figure 8. The instability of the electrical mode is only affected by the electric field, which is enhanced as the increases of Euler number. For the hydrodynamic mode, the instability is significantly influenced only when the electric field exceeds a threshold value.

In addition, the competition for dominant instability mode is also shown in figure 7. In most cases, the electrical mode is the dominant mode unless the Euler number is very small or large. In a weak electric field, the electrical mode will become the dominant mode as the flow is convectively unstable. However, the dominant mode would transform from the electrical mode to the hydrodynamic mode, with the increase of electric field in the absolute instability region.

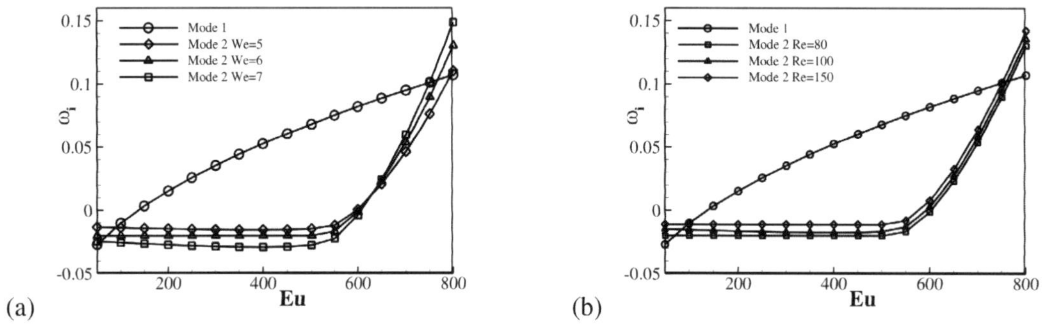

Figure 7. The absolute growth rates of the two modes versus the Euler number. (a) When Re = 80, (b) When We = 6.

For the hydrodynamic mode, larger Weber numbers weaken the instability in a weak electric field, but intensify the instability in a strong electric field, implying that surface tensor has both destabilizing and stabilizing effects. The increase of Reynolds number enhances the instability of the hydrodynamic mode indicating that viscosity has a destabilizing effect. In terms of dominant mode competition, the increase of Weber number promotes the transition to occur under a smaller Euler number, favoring the hydrodynamic mode to be dominant, but the increase of Reynolds number has a negative effect on the transition.

The AI-CI transition boundary of the two modes on Re −We plane is separately shown in figure 8. The boundary of electrical mode exists only when it is under the AI-CI critical Euler number Eu = 137. The upper right region is the CI region and the lower left region is the AI region. For the hydrodynamic mode, the left region is the CI region, and the right region is the AI region. As the electric field increasing, the boundary moves left and tends to be a vertical line, which indicates that the AI region is enlarged and its instability becomes less affected by Reynolds number.

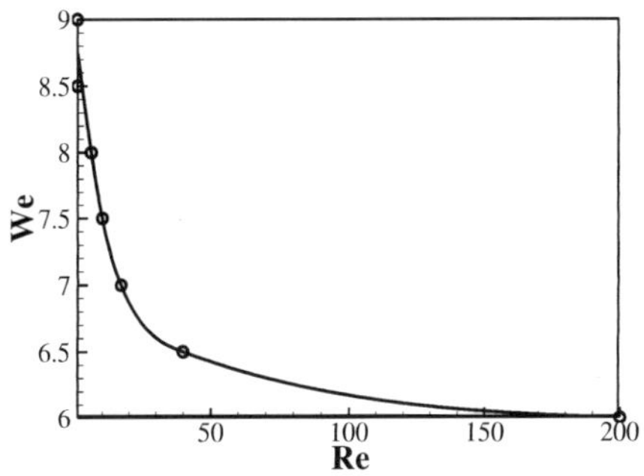

(a)

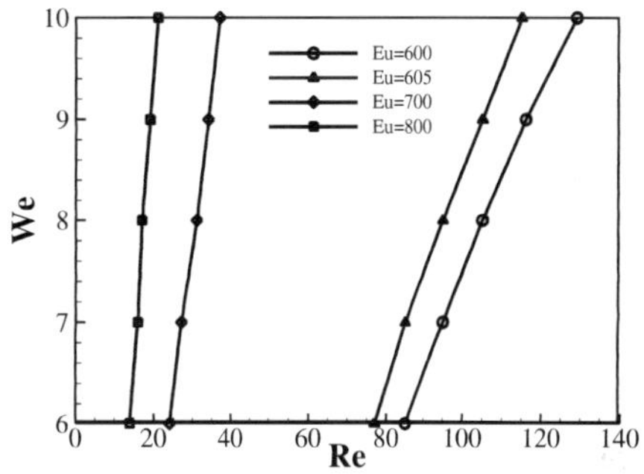

(b)

Figure 8. The AI-CI transition boundary of the two modes. (a) The electrical mode under Eu=137. (The CI region is in upper right, the AI region is in lower left region) (b) The hydrodynamic mode. (the CI region is in left, and the AI region is in right).

In summary, the spatial-temporal instability behavior of an electrified viscoelastic liquid sheet is investigated. A new long-wave unstable mode is found, besides a hydrodynamic mode. For the electric mode, its instability only depends on the electric field. And only when Eu = 137, which is the critical Euler number, Weber number and Reynolds number play roles in the AI-CI transition. For the hydrodynamic mode, the Euler number is able to affect the instability only when it exceeds a threshold value. Surface tensor shows destabilizing and stabilizing effects, while viscosity shows a destabilizing effect. The dominant instability mode mainly depends on the electric field, but the Weber number and Reynolds number can play roles in the mode competition.

References

[1] R. Q. Twiss, On oscillations in electron streams, *Proceedings of the Physical Society*. Section B **64** (8) (1951) 654
[2] P. Huerre, P. A. Monkewitz, Absolute and convective instabilities in free shear layers, *Journal of Fluid Mechanics* **159** (1985) 151–168.
[3] P. Huerre, P. A. Monkewitz, Local and global instabilities in spatially developing flows, *Annual Review of Fluid Mechanics* **22** (1990) 473–537. 120
[4] R. Briggs, Electron-stream interaction with plasmas, res. monogr., 29 (1964).
[5] A. Bers, Theory of absolute and convective instabilities, *Inst. Theoretical Physics*, 1973, p. a2/1 pp., International Congress on Waves and Instabilities in Plasmas, 2-7 April 1973, Innsbruck, Austria
[6] Barlow, N.S., Helenbrook, B.T., Lin, S.P., 2011. Transience to instability in a liquid sheet. *J. Fluid Mech.* 666, 358–390
[7] Yang, L., Fu, Q., 2012. Stability of confined gas liquid shear flows in recessed shear coaxial injectors. *J. Propul. Power.* 28, 1413–1424.
[8] M. F. El-Sayed, M. I. Syam, Electrohydrodynamic instability of a dielectric compressible liquid sheet streaming into an ambient stationary compressible gas, *Archive of Applied Mechanics* **77** (2007) 613–626.

[9] L.-j. Yang, Y.-x. Liu, Q.-f. Yin, Fu, Linear Stability Analysis of an Electrified Viscoelastic Liquid Jet, *Journal of Fluids Engineering* **134** (2012) 160 1024–1032.
[10] A.-C. Ruo, K.-H. Chen, M.-H. Chang, F. Chen, Instability of a charged non-Newtonian liquid jet, *PHYSICAL REVIEW E* 85 (1, 2).
[11] Lujia Liu, Yingjie Liu, Lipeng Lu, 2015, Atomization mechanism of a charged viscoelastic liquid sheet, *Chinese Journal of Aeronautics*, Volume 28 (2), 403-409
[12] Marco L, 2015, Different regimes of the uniaxial elongation of electrically charged viscoelastic jets due to dissipative air drag, *Mechanics Research Communications*, 69, 97-102
[13] Tong, MX; Fu, Qing-Fei; Yang, Li-Jun; two-dimensional instability response of an electrified viscoelastic planar liquid sheet subjected to unrelaxed axial elastic tension, *Atomization and Sprays* , 2015, Vol.25(2), p.99-121

IOP Conf. Series: Materials Science and Engineering **222** (2017) 012013 doi:10.1088/1757-899X/222/1/012013

Co_3O_4-Carbon Cloth free standing cathode for lithium sulfur battery

Jing Xu, Dawei Su, Guoxiu Wang*

Centre for Clean Energy Technology, University of Technology Sydney, NSW 2007, Australia.

*E-mail: Jing. Xu@student.uts.edu.au

Abstract. Lithium–sulfur (Li–S) battery has been considered to be one of the most promising next-generation electrochemical energy-storage systems due to its high theoretical energy of 2600 Wh kg^{-1} with low cost. The insulating nature of both sulfur and the dissolution of polysulfides are the two primary challenges for the application of lithium sulfur batteries. Here, we developed a binder-free cathode by chemisorption of Co_3O_4 to carbon cloth (CC), which was used as a 3D current collector to accommodate a large amount of sulfur, multiwall carbon nanofiber (MWCNF) and carbon black (CB) hybrids within the conductive scaffold, enabling the fabrication of ultrahigh sulfur loaded electrodes. The interconnected carbon fibers established a long-range conductive matrix for an efficient electron transport, the multiple conductive pathways guarantee high sulfur utilization. More importantly, the high electrolyte absorbability of the Co_3O_4-CC-S current collector facilitates well-localized polysulfides within the Co_3O_4-CC-S, meanwhile, the polar Co_3O_4 could also effectively entrapped the intermediated polysulfides preventing their free diffusion to the lithium anode, guaranteeing good cycling stability. Consequently, the Co_3O_4-CC-S electrodes exhibit excellent electrochemical performance with sulfur loading of 4.3 mg cm^{-2}.

1. Introduction

We significantly stabilized cycle life of high sulfur loading binder-free cathode by chemisorption of Co_3O_4 to carbon fiber cloth, which was used as a 3D current collector to accommodate a large amount of sulfur, MWCNF and CB hybrids within the conductive scaffold, enabling the fabrication of ultrahigh sulfur loaded electrodes.[1,2] This special nanoarchitecture combines the advantage of strong chemisorption of lithium polysulfides as well as excellent electrical conductivity, enabling high sulfur utilization and effective trap of lithium polysulfides. When applied as cathode materials for lithium sulfur batteries, the cathodes exhibit a reversible capacity of 1007 mAh g^{-1} after 300 cycles.

2. Experiment

Firstly, Co_3O_4 nanocages were constructed on carbon cloth (CC) by a solvothermal method. In a typical procedure, 0.1 of poly(vinylpyrrolidone) (PVP, Mw = 360000 g mol^{-1}, 99 %) was well dispersed in 7.5 ml of deionized water and 7.5 ml of ethanol (95 %) by stirring treatment, then 0.5 mmol of $Co(NO_3)_2 \bullet 6H_2O$ (> 98%) were dissolved into the above dispersion to form a light pink solution by continuous stirring for 30 min. CC which was pre-treated by an acid treatment were performed to functionalize CC with oxygen containing groups. Pristine CC were cut into circular disks with a diameter of 1.4 cm and soaked in HNO_3 solution at 60 °C for 3 h. Acid-treated CC were then washed by water and dried at 60 °C for 24 h. Then transform the as-made solution and CC into a 25 ml

Teflon-Lined stainless steel autoclave and reacted at 180 °C for 12 h. When the reaction finished, the solution was cooled down to room temperature naturally and washed to obtain the black Co_3O_4-CC. Then heated at 450 °C for 1 h under N_2 atmosphere. Homogeneous sulfur-containing slurry was fabricated by mixing 90 wt% sulfur, 5 wt% carbon black, and 5 wt% MWCNFs in N-methyl-2-pyrrolidone (NMP) followed by high power ultrasonication for 0.5 h. The as-prepared Co_3O_4-CC was immersed into the slurry for 10 min, then the Co_3O_4-CC was removed from the slurry and placed in a vacuum oven at 60 °C overnight to obtain the Co_3O_4-CC-S electrode.

3. Results

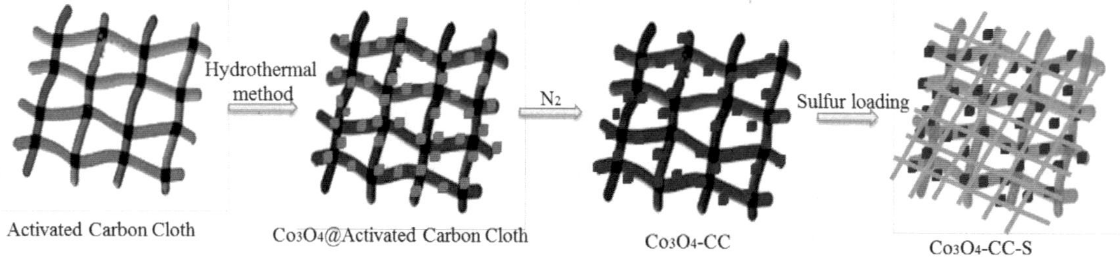

Scheme 1. Schematic illustration for the formation of Co3O4-CC-S.

By using our simple method, an activated carbon cloth was used as a template for the preparation of Co_3O_4-CC-S composite (Scheme 1), Co_3O_4 particles were successfully coated on the surface of CFCs. The as-prepared Co_3O_4 characterized by scanning electron microscopy (SEM) and transmission electron microscopy (TEM). As shown in Figure 1b, the as-made Co_3O_4 exhibits regular nanocubic morphology and has a uniform particle size of approximate 400 nm, also the surface of Co_3O_4 was smooth, which was consistent with the TEM result in Figure 1f, the surface of which shows high crystallinity without any detectable by-products as revealed by the X-ray diffraction pattern (Figure 2b). The structure of the Co_3O_4-CC-S was characterized by scanning electron microscopy (SEM) in Figure 1, it can be seen that sufficient void space, tens of micrometers in size, was generated by the randomly interconnected carbon fibers, which is capable of holding a large amount of active material, maintaining high electrolyte absorbability and effectively accommodating the volume expansion of sulfur during discharge.

The Co_3O_4-CC-S electrodes were prepared by immersing the Co_3O_4-CC in a premixed sulfur slurry containing 70 wt% commercial pure sulfur powder, 10 wt% carbon black(CB), and 20 wt% MWCNFs after high-power ultrasonication, it needs to be pointed out that MWCNFs were used in the sulfur slurry instead of conventional polymetric binders in integrated binder-free electrodes. From the X-ray diffraction pattern of the Co_3O_4-CC-S shown in Figure 2b, it can be seen that all the diffraction peaks are well indexed as sulfur structure, indicating that sulfur was successfully encapsulated into the porous structure. This seems to be further supported by the Raman spectroscopy analysis. In Figure 2c, we can detect distinct sulfur peaks at around 200 cm^{-1}, confirming that sulfur effectively diffuse into the Co_3O_4-CC composites. The 3D Co_3O_4-CC skelton is capable of holding a high, uniformly distributed concentration of the active material within its interconnected pores, enabling a high sulfur loading. We also characterized the structure of the sulfur-MWCNF-CB clusters loaded on the Co_3O_4-CC, considering the distribution of sulfur is closedly associated with the electrochemical performance of the electrode. It was seen that the MWCNFs and CB are not simply surrounded around the sulfur particles, but homogenously dispersed through the sulfur and formed an interconnected and embedded conductive network, the corresponding elemental maps of carbon and sulfur revealing that sulfur is uniformly distributed within the carbon conductive network constructed by MWCNFs and CB. In this work, we demonstrate that an embedded conductive scaffold formed by MWCNFs and CB can be constructed through the insulated sulfur, which is highly desirable for high sulfur utilization during the electrochemical redox process.

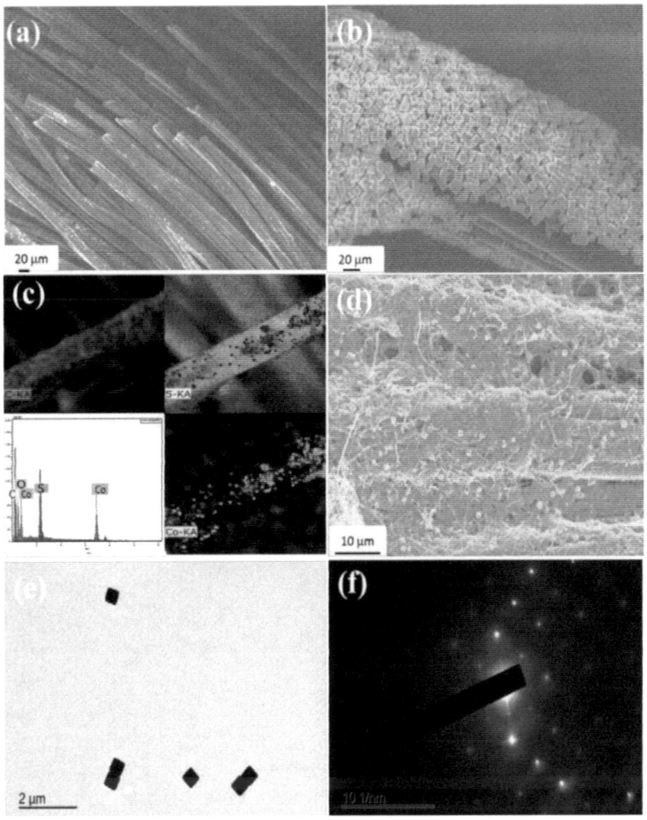

Figure 1. (a, b)SEM image of Co_3O_4-CC. (c)Mapping results of Co_3O_4-CC-S. (d)SEM image of Co_3O_4-CC-S. (e)TEM image of Co_3O_4. (f) SAED pattern for Co_3O_4.

Electrochemical measurements were conducted to compare the performance of Co_3O_4-CC-S and CC/S. Both the sulfur mass loading and the amount of electrolyte added in the coin cells were kept at the same level for Co_3O_4-CC-S and CC-S composites. Figure 3a, b illustrate the charge/discharge profiles of Co_3O_4-CC-S and CC/S in the first, second and 300th cycles, it is noteworthy that Co_3O_4-CC-S exhibits higher first-discharge plateau at around 2.37 V and wider second-discharge plateau at around 2.1 V compared with pure CNF/S electrodes, and even after 300 cycles, the discharge plateau still remain well, indicating more sulfur utilization in the Co_3O_4-CC-S. The cycling performance and Coulombic efficiency of the Co_3O_4-CC-S composite and pure CC-S electrodes over long-term cycles are shown in Figure 3d. After 300 cycles, the reversible capacity of Co_3O_4-CC-S is around 1007 mAh g^{-1}, the capacity retained 81% of the initial capacity after 300 cycles at 0.5C, while the stability of pure CNF/S electrode is significantly lower than that of Co_3O_4-CC-S, the capacity degraded to merely 434 mAh g^{-1} by 300th cycles. We also cycled the Co_3O_4-CC-S electrodes at different current rates. Figure 3c shows the results of Co_3O_4-CC-S and CC-S electrode cycled at step-wise current rates. During the discharge-charge process, the electrodes were consecutively cycled at 0.1C, 0.5C, 1C, 2C, 3C and then reversed back to low rates. It should be noted that when the current reversed back to 0.1C, the Co_3O_4-CC-S electrode could retain a capacity of 950 mAh g^{-1}, while the figure for CC-S electrodes was only 370 mAh g^{-1}.

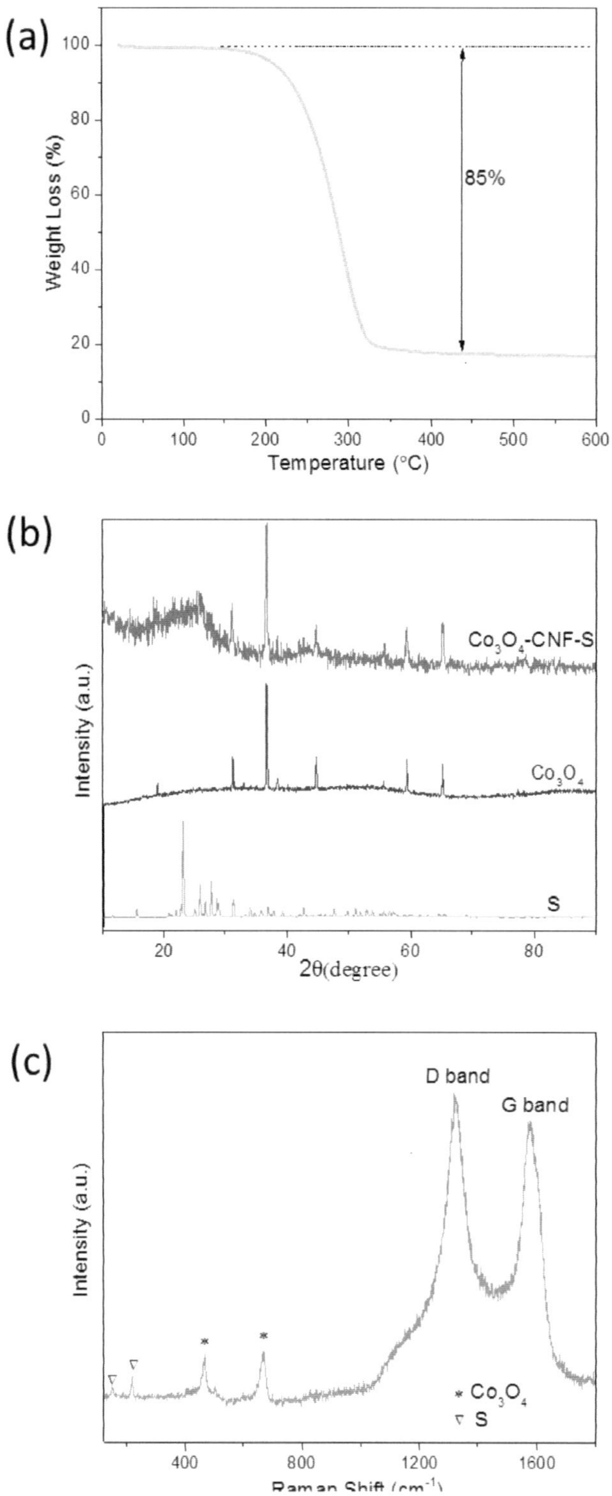

Figure 2. TGA curve(a), XRD patterns (b) and Raman spectra (c) of Co_3O_4-CC-S

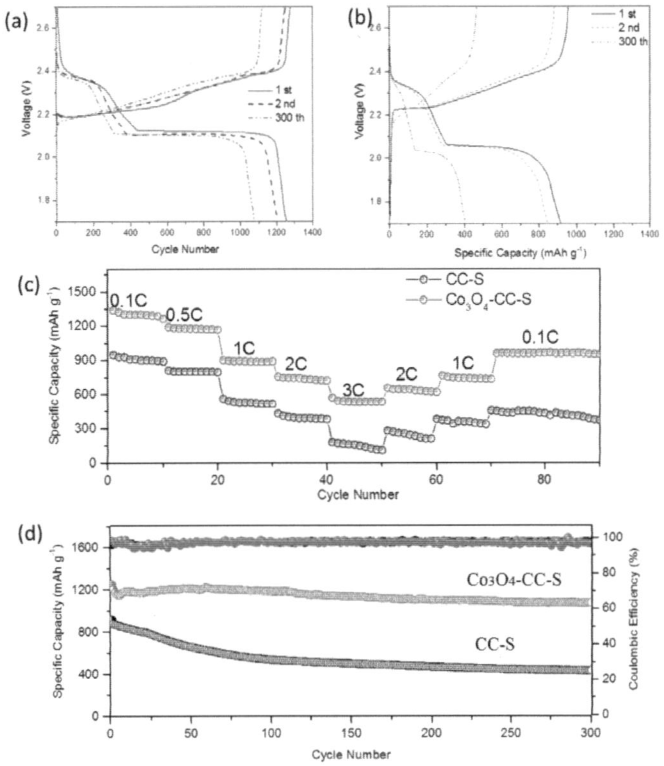

Figure 3. (a) Galvanostatic charge and discharge profiles of Co_3O_4-CC-S for different cycles at 0.2C. (b) Galvanostatic charge and discharge profiles of CC-S for different cycles at 0.2C. (c) Discharge/Charge capacity cycled at various rates from 0.1 C, 0.5 C, 1 C, 2 C, 5 C. (d) Long term cycling performance test of the Co_3O_4-CC-S, CC-S electrodes at 0.2C discharge rate and corresponding Coulombic efficiency.

4. Discusion

The interaction of different sulfur host materials (CC, Co_3O_4-CC) with lithium polysulfides were probed by visual discrimination, taking Li_2S_6 as the representative polysulfide. An equivalent amount of host materials were first added to Li_2S_6 solution. The superior intrinsic capability of Co_3O_4-CC to absorb Li_2S_6 was clearly obvious, as shown in Figure 4. The adding of Co_3O_4-CC rendered the Li_2S_6 solution light yellow after resting for 3h, implying strong adsorption while the carbon black solution did not exhibit obvious change, indicative of no interaction. Hence, the results from the visible experiment are highly consistent with the cycling performance of the batteries and further confirm the strong capability of Co_3O_4-CC-S in confining the LiPS species.

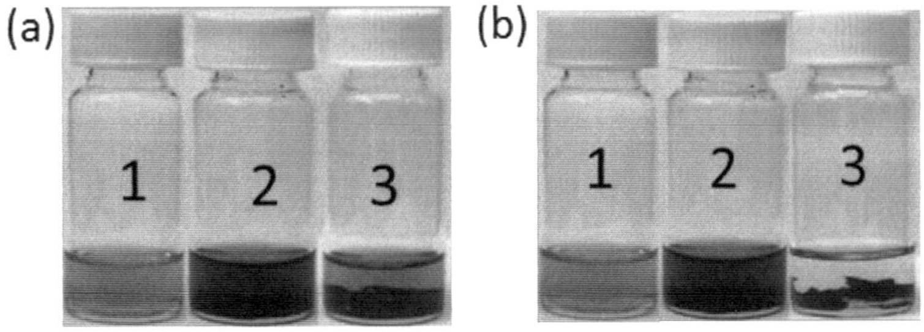

Figure 4. (a) Initially mixed. (b) Aging for 3h. ((1)L_2S_6 soultion. (2)CC-S adding into L_2S_6 soultion. (3) Co_3O_4-CC adding into L_2S_6 solution.)

5. Conclusion

In summary, we designed a binder-free cathode by chemisorption of Co_3O_4 to carbon fiber cloth, and applied this as a 3D current collector to accommodate a large amount of sulfur, multiwall carbon nanofiber (MWCNF) and carbon black (CB) hybrids within the conductive scaffold. The electrode with a high sulfur loading exhibits low polarization, stable cycling performance, and excellent rate capability compared with pure CC-S cathodes. The excellent performance could be attributed to synergic effects, including the high surface area and high conductivity of the CC-S matrix, as well as the strong polysulfides binding capability of Co_3O_4 particles. Moreover, Co_3O_4 nanocrystals probably participate in multiple polysulfide transformation, affecting the redox reaction environment favourably, leading to the improvement in the reversibility of Li_2S/polysulfides/S conversion. When applied as cathode material for lithium sulfur batteries, it exhibits superior cycling performance and excellent rate capability. This work opens a new opportunity for the realization of high-energy and commercially viable Li-S batteries.

References

[1] Fang R, Zhao S, Sun Z, Wang DW, Cheng HM, Li F. More Reliable Lithium-Sulfur Batteries: Status, Solutions and Prospects. *Advanced materials*, 2017.
[2] Wang J-G, Xie K, Wei B. Advanced engineering of nanostructured carbons for lithium–sulfur batteries. *Nano Energy* **15**, 2015.

IOP Conf. Series: Materials Science and Engineering **222** (2017) 012014 doi:10.1088/1757-899X/222/1/012014

Numerical simulation of hydrogen diffusion in the pressure vessel wall of a WWER-440 reactor

J. Toribio, D. Vergara and M. Lorenzo

Fracture and Structural Integrity Research Group (FSIRG),
University of Salamanca (USAL), Spain

Email:toribio@usal.es

Abstract. Materials forming the wall of a nuclear reactor pressure vessel (NRPV) can undergo in-service failure due to the presence of hydrogen, which enhances the fracture process known as hydrogen embrittlement (HE). A common way of avoiding this damage phenomenon is using a cladding material at the vessel wall side exposed to the hydrogenating source. This layer acts as a barrier for hydrogen diffusion and, hence, it protects the base material. In this paper, a numerical model of hydrogen diffusion *assisted by stress and strain* is used to analyse the hydrogen distribution, and hence the HE, in the pressure vessel wall of a real widely spread WWER-440 reactor considering two thickness for the cladding layer. Results show how the hydrogen accumulation is delayed as the thickness of the cladding layer increases, thus delaying the HE phenomenon affecting the structural integrity of the reactor.

1. Introduction

The wall of a nuclear reactor pressure vessel (NRPV) is composed by a bi-material of low carbon steel (base material) and stainless steel (cladding material) that form a cylindrical body of large dimensions ended by two half-spheres (figure 1). The aim of the second one, cladding material, is to protect the base material from the corrosion phenomena. The thickness of such a cladding layer is a key parameter for designing the NRPV wall and, therefore, an agreement between both cost and thickness must be reached. Nevertheless, the stainless steel is also affected by corrosion related phenomena [1] and the structural integrity of a NRPV under in-service conditions cannot be completely assessed. So, all the factors for contributing *hydrogen embrittlement* (HE) phenomena converge in a NRPV wall: (i) a hydrogenating environment, (ii) a material susceptible to HE, and (iii) a loading (mechanical or thermal) which generates a stress and strain state. During in-service conditions, a NRPV wall undergoes: (i) *thermal stresses* produced by the temperature gradient existing between the inner side (exposed to the hydrogenated source at high temperatures, T_i = 300 ºC) and the outer side (exposed to environment at room temperature, T_o = 25 ºC) and (ii) the *residual stresses* caused by the manufacturing process [2]. In [2] an estimation of the residual stress state caused by the heat treatment of a similar NRPV is revealed in terms of the tempering parameters: tempering time (t_{temp}) and tempering temperature (T_{temp}). From these residual stress distributions, the studies [3,4] revealed the key role of the global stress state, including such residual stress and plastic strain states and the thermal stress, on the HE process of the pressure vessel wall of a WWER-440 reactor. So, within this framework, the aim of this paper is to reveal if the currently used thickness of the cladding layer is enough for delaying both the hydrogen entry and the hydrogen accumulation in the base material layer, thereby assessing the safety operation of the NRPV against HE during their life in-service. To achieve this goal, different numerical simulations by finite element (FE) of the hydrogen diffusion assisted by stress and strain were carried out considering two cases of study in terms of the cladding layer thickness: the currently used one (8 mm) and a case considering a four times higher cladding thickness (32 mm). This way, the hydrogen distributions in a real NRPV wall (type WWR-440) are revealed in this paper and, consequently, the more adequate dimensions of the cladding layer can be stated for a safety operation of the NRPV.

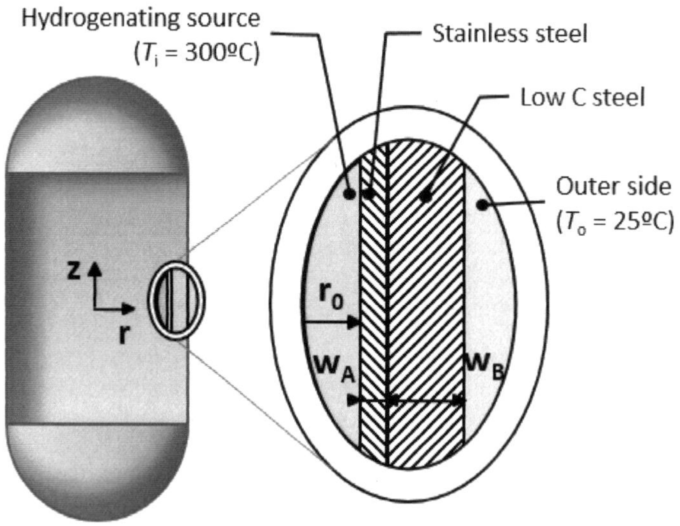

Figure 1. Scheme of a NRPV type WWER-440.

2. Stress and strain state

To reveal the hydrogen accumulation in the NRPV wall, a numerical simulation of the hydrogen diffusion assisted by stress and strain is carried out. To achieve this goal, it is necessary to determine the residual stress-strain state of the vessel wall during in-service conditions. In this study, the stress state was estimated by adding (i) the residual stress state obtained during the manufacture of the NRPV [2] considering the following tempering conditions (T_{temp} = 650 °C and t_{temp} = 100 h), and (ii) the stress caused by the thermal gradient during in-service conditions [3,4]: inner side T_i = 300 °C and outer side, room temperature T_o = 25 °C. The dimensions of a real *pressurized water reactor* (PWR) type WWER-440 are [2]: inner radius r_0 =1.77 m and a wall with two layers of thickness w_A =8 mm of stainless steel and w_B =142 mm of low carbon steel (figure 1). In this study, the HE of two different thicknesses for the cladding layer (w_A) were compared (table 1), the commonly used one (8 mm) and another four times higher ($4w_A$). For both cases, the same second layer thickness (w_B =142 mm) is considered in simulations.

Table 1. Dimensions of the constituent layers of a NRPV wall.

Case study	Wall A (w_A)	Wall B (w_B)	Whole wall (w)
I	w_A = 8 mm	142 mm	150 mm
II	$4w_A$ = 32 mm	142 mm	174 mm

The variables representing the stress and strain fields in the hydrogen diffusion model [5-8] used in this study are: the *hydrostatic stress* σ and the equivalent plastic *strain* ε^P. The distributions of such variables through the thickness *w* of the pressure vessel wall of a WWER-40 reactor are shown in figure 2. The distribution of the hydrostatic stress in the radial direction can be divided into three well defined zones (figure 2a): (i) *zone A* corresponding to the cladding layer; (ii) *zone* B^+ corresponding to the zone of the base material wall with tensile stresses, and (iii) *zone* B^- corresponding to the base material subjected to compressive stresses.

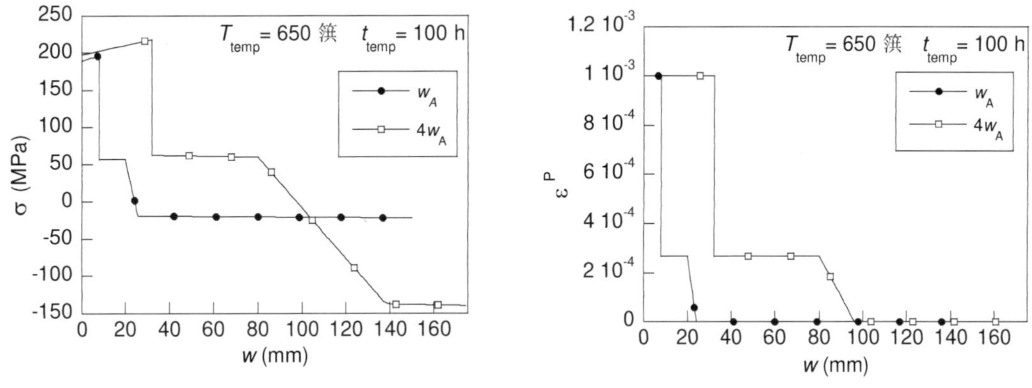

Figure 2. Distribution of (a) global hydrostatic stress σ, and (b) equivalent plastic strain ε^P, state through the NRPV thickness for the two cases of study.

3. Hydrogen embrittlement

A full description of the model of hydrogen diffusion assisted by stress and strain used in the numerical simulations of this paper is included in [5-8]. This way, figure 3 shows the distribution of the normalized hydrogen concentration (C/C_0, C_0 being the hydrogen concentration in a stress and strain free material) after the first year of service time (t_{serv}) of the PWR. The hydrogen concentration decreases with depth from the inner surface in contact to the hydrogenating environment. The cladding layer acts not only as a barrier against corrosion phenomena but it also acts as a delayer of hydrogen diffusion and accumulation in the NRPV wall. This way, an analysis of the cladding width as a *design parameter* against HE would allow a better statement of the structural integrity of a NRPV.

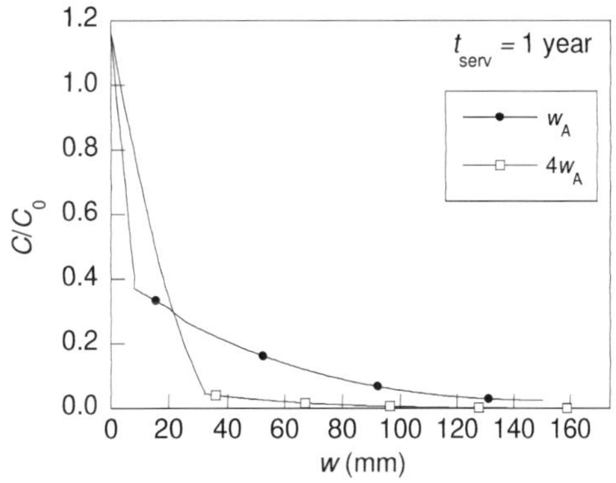

Figure 3. Distribution of the normalized hydrogen concentration through the vessel wall of a PWR after the first year of service time.

So, according to these results different behaviors against HE of a fission NRPV wall can be achieved just modifying the cladding thickness. This way, the optimum value for this parameter can be determined in order to assess the structural integrity of this component depending on the PWR service time (t_{serv}) and the supervision stops for maintenance. According to results, the real thickness of the stainless steel cladding layer (w_A) is not an absolute barrier against the hydrogen diffusion toward the base material of the vessel (highly susceptible to HE [9-11]). Therefore, the hydrogen concentration reached at critical points, such as interface, is not negligible, and consequently, a risk of catastrophic failure of the NRPV due to HE does exist.

4. Conclusions

The materials used in a nuclear reactor pressure vessel (NRPV), low carbon steel as base material and stainless steel for the cladding material, are both susceptible to hydrogen embrittlement phenomenon. The aim of the cladding layer is to act as a barrier for the hydrogen diffusion from the hydrogenating surface (inner side of the NRPV) towards the base material. In this paper, the hydrogen accumulation as result of the hydrogen diffusion assisted by stress and strain is analyzed for revealing if a commonly used cladding layer thickness is enough for delaying the hydrogen entry in the base material layer. Obtained results reveal that, for a realistic nuclear reactor service time (about 1 or 2 years), a cladding layer of stainless steel thicker than that commonly used nowadays in the nuclear industry should be recommended. It would reduce the hydrogen entry in the second vessel layer (base material of low carbon steel.

Thus, according to the numerical simulation carried out in this study, the currently used thickness of the cladding layer of the vessel of a pressurized water reactor type WWER-440 (8 mm of stainless steel) is *not* an absolute barrier against hydrogen diffusion towards the base material. According to the obtained results, a real barrier against hydrogen accumulation in the base material layer is achieved considering a cladding layer of thickness four times higher than the one used is recommended. This way, the following aspects seem to be convenient for obtaining a real protection of the base material layer against the harmful effect of hydrogen: (i) the selection of an adequate material to be used in the cladding layer and (ii) the analysis of hydrogen diffusion assisted by stress and strain in the NRPV to guarantee the efficiency of the cladding layer. Thus, such aspects could be considered as future research to be developed.

References

[1] Louthan MR and Derrick RG 1975 *Corros. Sci.* **15** 565
[2] Kostylev VI and Margoli BZ 2000 *Int. J. Pres. Ves. Pip.* **77** 723
[3] Toribio J, Vergara D and Lorenzo M 2016 *Proc. Struct. Integrity* **2** 622
[4] Toribio J and Vergara D and Lorenzo M 2017 *Energy Proc.* [in press]
[5] Toribio J and Kharin V 1998 *Nucl. Eng. Des.* **182** 149
[6] Toribio J and Kharin V 2000 *Fusion Eng. Des.* **51-52** 213
[7] Toribio J, Kharin V, Vergara D and Lorenzo M 2010 *Adv. Mater. Res.* **138** 117
[8] Toribio J and Kharin V 2015 *Philos. Mag.* **95** 3429
[9] Toshimitsu A, Takenao JR, Satoh K and Yamada T 1996 *Eng. Fract. Mech.* **1** 47
[10] Pound BG 1998 *Acta Mater.* **16** 5733
[11] Hadam U and Zakroczymski T 2009 *Inter. J. Hydrogen Energ.* **34** 2449

IOP Conf. Series: Materials Science and Engineering **222** (2017) 012015 doi:10.1088/1757-899X/222/1/012015

Techno-Economic and dynamic analysis of low velocity wind turbines for rural electrification in agricultural area of Ratchaburi Province, Thailand

Sumate Lipirodjanapong[1,a], Weerapol Namboonruang[2,b]

[1]Department of Electrical Technology , Faculty of Science and Technology, Muban Chombueng Rajabhat University, (MCRU), Thailand
[2] Department of Civil Technology and Architecture , Faculty of Science and Technology, Muban Chombueng Rajabhat University, (MCRU), Thailand

Email: [a]LSUMATE@mcru.mail.go.th , [b]bannork_civil@hotmail.com

Abstract. This paper presents the analysis of potential wind speed of electrical power generating using for agriculture in Ratchaburi province, Thailand. The total area is 1,900 square kilometers. First of all, the agriculture electrical load (AEL) data was investigated from all farming districts in Ratchaburi. Subsequently, the load data was analyzed and classified by the load power and energy consumption at individual district. The wind turbine generator (WTG) at capacity rate of 200w, 500w, 1,000w, and 2,000w were adopted to implement for the AEL in each area at wind speed range of 3 to 6 m/s. This paper shows the approach based on the wind speed at individual district to determine the capacity of WTG using the capacitor factor (CF) and the cost of energy (COE) in baht per unit under different WTG value rates. Ten locations for wind station installations are practical investigated. Results show that for instance, the Damnoen Sa-duak (DN-04) one of WTG candidate site is identically significant for economic investment of installing rated WTG. The results of COE are important to determine whether a wind site is good or not.

1. Introduction

Generally, among various renewable energy resources, wind power energy is one of the most popular and promising energy resources in the whole world. With the specific wind turbine, the available electricity generated by a wind power generation system depends on mean wind speed (MWS), standard deviation of wind speed and location of installation [1]. The importance of utilizing the renewable energy system, wind turbine generator (WTG) has been attracted greatly in these days because the agriculture electrical load demand is need for the renewable energy resource and it has formulated as a national strategy for the development of renewable energy application and energy conservation measures. For this paper purpose, continues an effort to study more wind potential on individual distinct (10 districts) of Ratchaburi province in Thailand with optimal cost and number of WTG installation in area-base. Today's wind turbines are state-of-the-art-of modern technology modular and very quick to install. Stand-alone power generation system is an important alternative source of electrical energy, find applications in location where convention generation is not practical. Consider, for instance, agriculture farm or ranches located far away from main power lines. It has been shown that a remote load has only to be a matter of a few kilometer away from a main power transmission line for a stand-alone wind turbine generator to be cost-effective [2],[3] and cornerstone in the future green energy market [4],[5],[6].

This paper focuses on the agriculture electrical load (AEL) to determine the capacity of WTG under rated wind speed data recorded from area-base location in Ratchaburi province of Thailand. The cost of energy (COE), capacity factor (CF) and capital costs of WTG under different location are also determined. These parameters can be used to evaluate the performance of wind speed and wind turbine generator installation in each distinct of Ratchburi province.

2. Methodology

In this paper, the economic analysis of the electrical power generating from low speed wind range about 3 to 6 m/s was investigated. There are the procedures of calculation as following, the investigation of wind speed data and AEL in area-base. Then, the numbers of WTG installation have calculated by Eq. 1 by the individual AEL, can be as express

$$WTG_{RATED} = \left(\frac{Energy_{USED}}{Capacity_{hour}} \right) \times 1.25 \tag{1}$$

Where $Energy_{USED}$ is the used energy in watt-hour unit, and $capacity_{hour}$ is the ability of WTG. It can be generated the electrical energy (watt-hour) in period 8 hour/day and the capacity factor (CF) can be calculated by Eq.2

$$CF = \frac{L_{avg}(kW)}{(kW_{inst})} = \frac{kWh_{gen}(year)}{(kW_{inst}) \times 8{,}760} \tag{2}$$

Where $Lavg_{(kW)}$ is the average load, kW_{inst} is the power generating installation, kWh_{gen} is the electrical energy generating per year, and kW_{inst} is the electrical energy installation per year (hour).

The cost of energy [7] is free of detailed economic variables. The term COE is defined as the unit cost produce energy (in baht per kilowatt-hour) from WTG [8], and it can be expressed by Eq. 3

$$COE = \frac{NPV(C_A)}{E_I} = \frac{C_I}{(8{,}760 \times n)} \times \left(\frac{1}{P_R C_F} \right) \times \left\{ 1 + m \left(\frac{(1+I)^n - 1}{I(1+I)^n} \right) \right\} \tag{3}$$

Where C_i is the capital cost in baht, P_R is the power generating rate, C_F is the capacity factor, m is the average annual operation and maintenance cost in baht, I is the interest rate, and n is the lifetime of WTG.

3. Investigation of agriculture electrical load data

The Agriculture electrical load was investigated in Ratchaburi area by the collection of information outside a laboratory. The Ratchaburi is a province, which is consist of ten districts. From the AEL data, it can be clasified device into 2 groups that are water pump into the farm and electrical lamp, fan etc. The total agriculture elecrical load of each district is shown table 1.

Table 1 Amount of using Electric in area-base Agriculture Self-Reliance in Ratchaburi Province

Area-base in Ratchaburi (10 districts)	Amount of using Electric in area-base Agriculture Self-Reliance in Ratchaburi Province					
	AEL in area-base, (kW)	Electrical energy in area-base (kWh)	Area-base for case study (m^2)	Expected energy supplied for load (kWh/m^2)	All agriculture area-base (Sq.km)	Expected energy supplied for all load (kWh)
Mueang Ratchaburi (M.-01)	1.026	5.626	3,600	2.813	188.35	331,149
Wat Phae (WP.B-02)	1.865	2.984	6,400	0.746	34.55	16,108
Damnoen Saduak (DN.-04)	5.222	28.348	3,600	14.174	161.52	1,430,879
Pak Tho (PT.-08)	2.707	10.540	14,400	1.171	261.44	191,338
Photharam (PTR.-05)	2.338	28.056	8,000	5.611	227.41	797,514
Bang Phae (BPH.-03)	4.476	89.52	12,800	11.19	111.92	782,707
Ban Pong (BP.-06)	2.707	0.756	9,600	0.126	202.05	15,911
Chom Bueng (CH.B-07)	1.227	2.346	4,800	0.782	359.76	175,833
Ban KhaBK.-10	1.624	3.614	8,000	0.628	216.55	84,994
Suan PhuengSP-09	1.937	2.827	16,000	0.283	144.48	25,555
	Total					3,851,988

4. Wind turbine power model

A typical configuration of permanence magnet synchronous generator (PMSGs) bases wind energy conversion system consists of a generator side converter interconnected through an intermediate DC-link as shown in figure 1 which Given a plant size in terms of annual energy production, the COE can be minimized by choosing the optimum GRR [9].

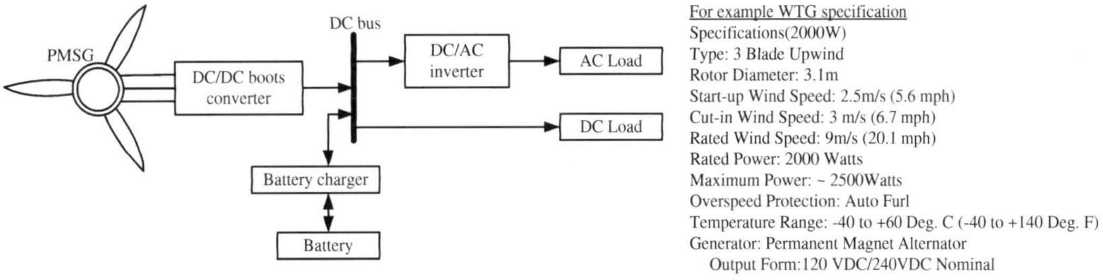

Figure1. Typical configuration of wind energy conversion system for agriculture electrical load

In this paper, the 4 rate types of WTG such as 200w, 500w, 1,000w and 2,000w were studied. Result is shown in figure 2

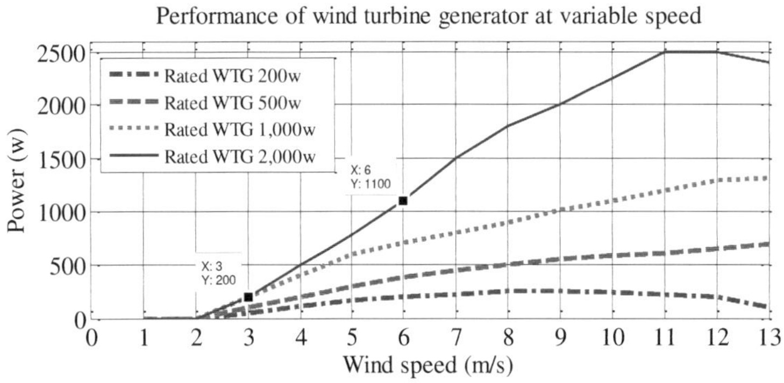

Figures 2. Power and Wind Speed of wind turbine generator (PMSG) and practical wind speed test

Figure 2 shows the rated wind turbine generating are differently 4 class criteria. Rated WTG 2000w is the maximum power generator and 200w is the minimum power generator. Where the blue line is represented the maximum WTG at rated 2000w, considering speed range 3 to 6 m/s therefore at the 3 m/s is cut-in wind speed for starting up operating WTG point. The purple line, the green line and the red line are represented the rated WTG 1,000w, 500w and 200w respectively. Figure 2 (in the right hand side picture) shows the practical wind speed test in Damnoen Saduak (DN-04) site at the tower height of 8 meter. It can collect data at the average wind speed rage from 3 to 6 m/s.

5. Economic analysis of wind power generating

Form the Table 1 is the total agriculture electrical loads; it has obtained from the fieldwork. The AEL data is calculated by Eq.1 and it obtained the results that are the rated wind turbine generator and number installation. From these results, it can be evaluated the economic investment that is CF and COE.

For example, Damnoen Sa-duak district, it is found that a total agriculture electrical load is 1,632 kilowatt/sq.km and all electrical energy consumption is 8,858.75 kilowatt-hour/sq.km. And the total energy can be calculated from the total power at capacity WTG in 8 hour. The capital cost was estimated by market prices in Thailand. The CF of power wind generating is the ratio of its actual output over a period of time. So, the COE can be calculated from net present value (NPV) with consideration of the interest rate 12% and life time of WTG 20 years. Results are shown in Table 2 to 5.

Tables 2 Indies Performance of WTG vary from the Fluctuating Wind Speed (200w)

Wind Speed (m/s)	WTG 200w	Number wind Site (Station)	Total Power (Watt)	Total Energy (Whr) at 8hr.	Capital Cost* (Baht)	CF	COE** (baht/unit)
3	50	45	2,250	18,000	742,500	0.0833	38.0632
4	112	20	2,240	17,920	330,000	0.1867	7.5859
5	166	14	2,324	18,592	231,000	0.2767	3.4532
6	200	11	2,200	17,600	181,500	0.3333	2.3789

*included the all converter and battery system price.
**life time of WTG about 20 year and interest rate 12% per year

Tables 3 Indies performance of WTG vary from the variable wind speed (500w)

Wind Speed (m/s)	WTG 500w	Number wind Site (Station)	Total Power (Watt)	Total Energy (Whr) 8hr.	Capital Cost* (Baht)	CF	COE** (baht/unit)
3	100	22	2,200	17,600	803,000	0.0667	39.4197
4	205	11	2,255	18,040	401,500	0.1367	9.3801
5	300	8	2,400	19,200	292,000	0.2000	4.3800
6	395	6	2,370	18,960	219,000	0.2633	2.5265

*included the all converter and battery system price.
**life time of WTG about 20 year and interest rate 12% per year

Tables 4 Indies performance of WTG vary from the variable wind speed (1,000w)

Wind Speed (m/s)	WTG 1,000w	Number wind Site (Station)	Total Power (Watt)	Total Energy (Whr) 8hr.	Capital Cost* (Baht)	CF	COE** (baht/unit)
3	200	11	2,200	17,600	924,000	0.0667	45.3597
4	400	6	2,400	19,200	504,000	0.1333	22.6798
5	600	4	2,400	19,200	336,000	0.2000	15.1199
6	700	3	2,100	16,800	252,000	0.2333	12.9599

*included the all converter and battery system price.
**life time of WTG about 20 year and interest rate 12% per year

Tables 5 Indies performance of WTG vary from the variable wind speed (2,000w)

Wind Speed (m/s)	WTG 2,000w	Number wind Site (Station)	Total Power (Watt)	Total Energy (Whr) 8hr.	Capital Cost* (Baht)	CF	COE** (baht/unit)
3	200	11	2200	17,600	1,100,000	0.0333	107.9992
4	500	5	2500	20,000	500,000	0.0738	22.0189
5	780	3	2340	18,720	300,000	0.1300	7.1005
6	1,100	2	2200	17,600	200,000	0.1833	3.5702

*included the all converter and battery system price.
**life time of WTG about 20 year and interest rate 12% per year

From the table 5, for instance, at wind speed of 3 m/s, rated wind turbine generator 2,000w can generate the total electrical power and the total electrical energy which is about 2,200w and 17,600 Whr respectively. This electrical energy value was evaluated by the capacity of WTG in period 8 hour per day (24hour). The number wind site, 11 sites is calculated from Eq.1 which considered in term of the used energy and capacity of WTG at wind power generating of variable wind speed. Both capacity factor (CF) and cost of energy (COE) can be calculated from Eq. 2 and Eq. 3 that are equal to 0.0333 and 107.9992 baht per unit respectively. From the results, the COE depended upon the CF value. When the CF value is increasing affect to the COE is decreasing value. That is relation of between the COE and CF is inverse function. It can conclude that it is suitable number of install WTGs and economic wind power generating.

6. Result

All agriculture electrical loads installed in the Damnoen Saduak distric are calculated by equation (1), (2), and (3). The results show the indices of capacity factor, cost of energy, wind power output (watt), electrical energy (watt-hour) and cost of investment per wind speeds (3 to 6m/s) as shown in figure 3

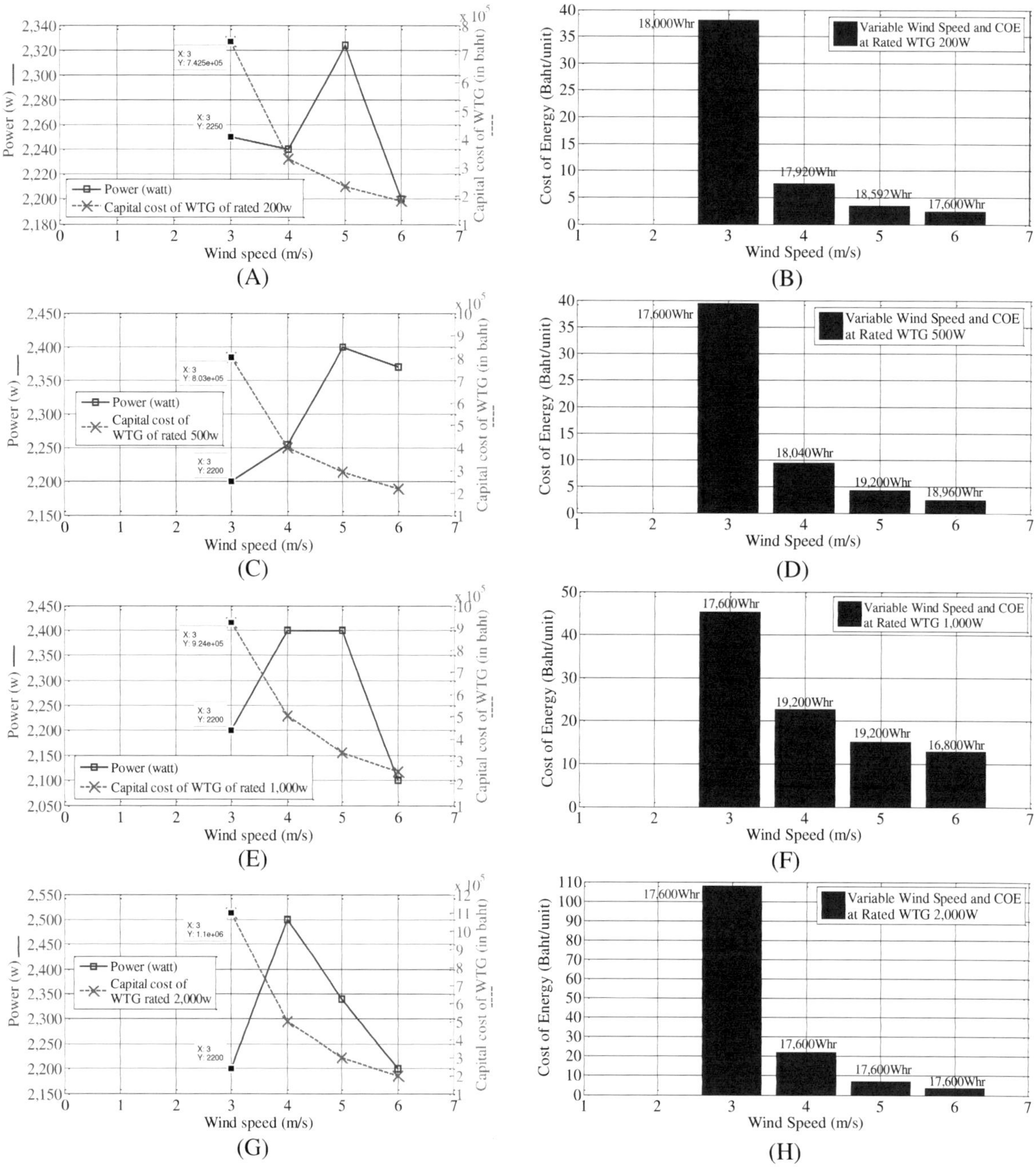

Figures 3 Result of Performance Wind Turbine Generator and Cost of Energy at 3-6 m/s

(A-C-E-G) Performance Power Generating and Invesment of WGT

(B-D-F-H) Economic of Electrical Energy Generating (baht per unit)

From figure 3 (in the left hand side) show the relationship of the cost of investment and the wind power generating rates at 200w, 500w, 1,000w and 2,000w (A-C-E-G) respectively. All results are considered at the wind speed rage of 3 to 6 m/s. Where the dash-blue line represents the cost of investment of WTGs install (baht). The dense rad line represents the wind power generating output (watt) from variable wind speeds.

In the right hand side, figure 3 show the relations of the cost of energy in baht per unit and capacity of wind electrical energy generating at vary wind speed range of 3 to 6 m/s. The wind power generators were implemented at rate of 200w, 500w, 1,000w and 2,000w (B-D-F-H) respectively. Results show that with increasing of the wind speed, the cost of energy and amount of WTG installation significantly decrease.

7. Summary

In this paper, the analysis of the average wind speed potential in the west region of Thailand in Ratchaburi province, which consists of ten districts, is studied. The individual districts are investigated the AEL data for the evaluation of economic investment of WTG installation. The rated WTG have adopted to install at 200w, 500w, 1,000w and 2,000w respectively. From the AEL data, the average wind speed at individual district are determined the capacity of WTG using capacitor factor and cost of energy in baht per unit under different values of rated WTG. Results show that with increasing of the wind speed, the cost of energy and amount of WTG installation significantly decrease. These results can be the information for the government to decide the investment of WTG installation.

References

[1] M. Sathyajith, "Wind Energy: Fundamentals, Resource Analysis and Economics," (2006).

[2] W. D. Kellogg, M. H. Nehrir, G. Venkataramanan, and V. Gerez, "Generation unit sizing and cost analysis for stand-alone wind, photovoltaic, and hybrid wind/PV systems," *IEEE Trans. Energy Conversion.*, vol. 13, no.1, pp. 70-75, (Mar. 1998).

[3] T. Sejyu, T. Nkaji, K. Uezato and T. Funabashi, "A hybrid System Using Alternative Energy Facilities in Isolated Island," *IEEE trans. Energy Conversion*, vol. 20, no. 2, pp. 406-414, (Jane 2005)

[4] D. Carvalho, A. Rocha, M. G_omez-Gesteira, C. Silva Santos, Potential impacts of climate change on European wind energy resource under the CMIP5 future climate projections, *Renew. Energy* 101 (2017) 29-40.

[5] H.-G. Kim, H.-J. Hwang, S.-H. Lee, H.-W. Lee, Evaluation of SAR wind retrieval algorithms in offshore areas of the Korean Peninsula, *Renew. Energy* 65 (2014) 161-168.

[6] C. Mattar, D. Borvar_an, Offshore wind power simulation by using WRF in the central coast of Chile, *Renew. Energy* 94 (2016) 22-31.

[7] J. F. Manwell, J. G. Mcgowan, and A. L. Rogers, Wind Energy Explained Theory, *Design and Application*. New York: Wiley, (2002).

[8] Maria Isabel Blanco, The Economics of Wind Energy. *Renewable and Sustainable Energy Reviews* 13 (2009) 1372-82.

[9] Matthiias Preindl, Silverio Bolognani, Optimization of the Generator to Rotor Ration of MW Wind Turbine based on the Cost of Energy with Focus on Low Wins Speeds, *International conference on IEEE Industrial Electronics Society (IECON2011)*, pp. 906-911.

IOP Conf. Series: Materials Science and Engineering **222** (2017) 012016 doi:10.1088/1757-899X/222/1/012016

Optimization of woven jute/glass fibre-reinforced polyester hybrid composite solar parabolic trough collector

K S Reddy[1] and Hitesh Singla[1]

[1]Heat Transfer and Thermal Power Laboratory, Department of Mechanical Engineering, Indian Institute of Technology Madras, Chennai-600036, India.

Email: hiteshsingla90@gmail.com

Abstract. In the present work, structural analysis of 5.77m × 4m woven jute (J)/glass (G) fibre-reinforced polyester hybrid composite solar parabolic trough is carried out based on trough parameters to obtain the minimum RMS local slope deviation, termed as SDx value under gravity loading. The optimization is done by varying parameters viz. direction and size of reinforced conduits, stacking number and sequence of hybrid trough laminate at fibre orientation of $\Delta\theta$=45° and $\Delta\theta$=60° amongst the layers at 0° collector angle. The analysis revealed that the configuration in which the conduits are placed in both X and Y directions is preferred over other configurations to scale down the effect of wind loads. Furthermore it has been observed that laminate of the order $[0°_G/45°_G/-45°_J/90°_J]_s$ undergoes minimum surface deformation amongst all the other configurations at conduit reinforcement in both X and Y directions for a conduit thickness of 0.75 mm and radius of 10 mm and obtains the overall SDx value of 1.3492 mrad. The results shows that proposed trough model is very promising and evolves a cost effective system.

1. Introduction

The effective and cost-effective use of solar energy is of paramount importance when it comes to using alternate sources of energy. Over the past few years, the researchers have tried to analyse potential solar devices by making use of new materials and studying them from structural, thermal and optical outlook. Hence, the composite materials are gaining momentum to make solar collector systems and one such use is in solar parabolic trough based technology, in which trough reflectors concentrates the direct solar radiations onto an absorber tube located in the focal line of parabola [1]. The amount of radiations reflected from the trough reflector that are actually falling on the absorber tube depends upon collector's strength to maintain its ideal shape under gravity and wind loads. Thus, the need of using superior materials arises to achieve higher optical efficiency at low manufacturing costs. Natural fibre polymer composite materials are one such class of materials that are cost effective, have low density and offers sufficiently high specific strength. The attempts have been made in the recent years to fabricate the composite trough but still they are not used predominantly due to high costs involved in making its mould. The use of polyester resin and woven fibre glass cloth for parabolic trough after doing reinforcement with PVC conduits in both its transverse and axial direction by Kalogirou et al. [1], the use of polyester resin and chopped stranded fibre glass cloth after placing the PVC conduits in only longitudinal direction by Valan Arasu and Sornakumar [2], the use of epoxy woven fibre glass composite trough by doing reinforcement with polystyrene strips by Gianlucca

Coccia et al. [3] are few of the examples. But all these configurations saw the use of synthetic fibres in an experimental work.

For obtaining the cost effective and high stiffness/weight ratio structure, there has to be a use of alternate materials, their repair and its optimization [4]. Therefore, in the present work, the use of Woven/Jute Glass Fibre Reinforced Polyester Hybrid Composite (WFRPH) material in Parabolic Trough Collector (PTC) is exercised on the basis of stiffness/weight and cost/performance of the trough. After selecting the material, its optimization is done by varying the parameters like direction and size of conduits, stacking number and sequence of hybrid Trough Laminate (TL) at different fibre orientations and then is analysed structurally at 0° collector angle under gravity load.

2. Methodology

2.1. Geometrical modelling and numerical procedure

A WFRPH PTC consists of two columns of composite mirrors having total of 8 mirrors and each having dimensions of 1439.5 × 1998 × 4 mm that are attached to the support structure as shown in figure 1. The composite mirrors or the trough is made of stacks of laminae with direction specific fibres to form the TL which is reinforced with PVC conduits in both X and Y directions to increase its stiffness/weight ratio. Each mirror is separated from its adjacent mirror by a gap of 4 mm. The support structure consists of 8 arms that are tied to the trough at specific points and the mirror arms are connected to the pylons through torque tube as shown in figure 2.

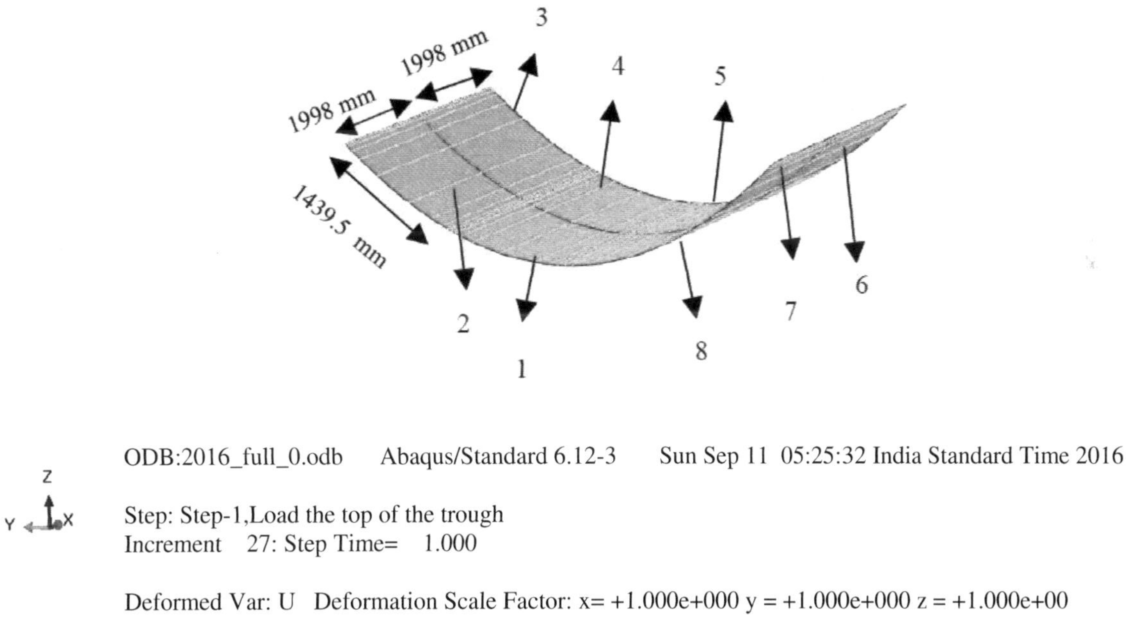

Figure 1. WFRPH Parabolic trough reinforced with PVC conduits with mirrors numbered from 1-8.

IOP Conf. Series: Materials Science and Engineering **222** (2017) 012016 doi:10.1088/1757-899X/222/1/012016

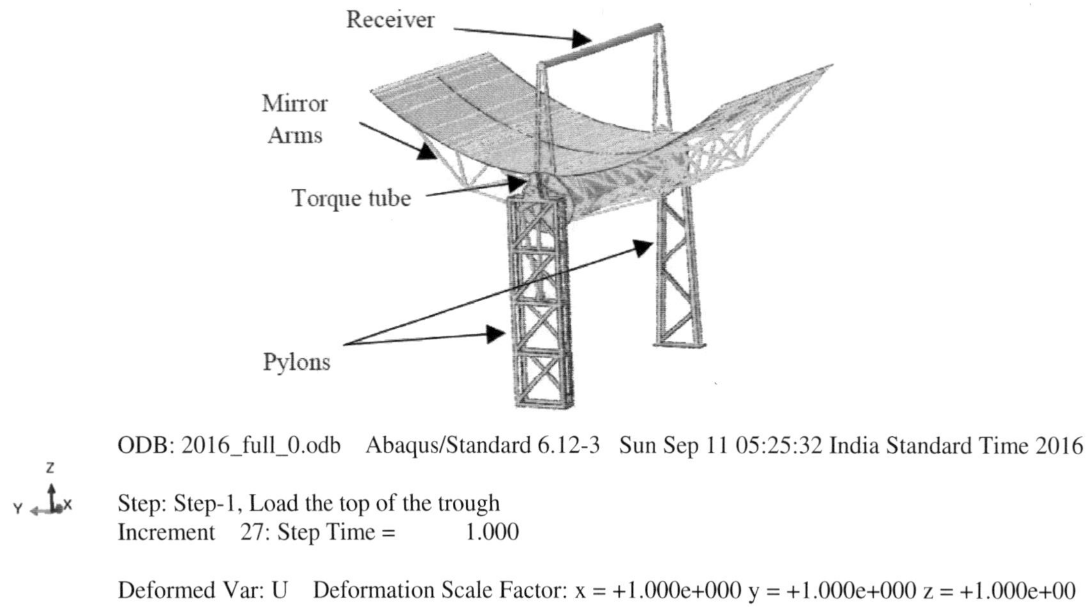

Figure 2. A WFRPH solar PTC.

Following mesh element types are used to ensure convergence and mesh element size and total number of elements are found to be 20 mm and 3,05,592 respectively after mesh convergence study.

a) Trough: S8R (eight noded quadrilateral shell element with reduced integration)
b) Arm, Receiver tube, Torque Tube & Pylons: C3D10, C3D15, C3D20R (Quadratic tetrahedral element, quadratic wedge & quadratic hexahedral element respectively)
c) Conduits used for reinforcement: B31 (linear line element)

The basic formulation of equilibrium equations is done using Newton's full solution technique for the finite element model [5].

$$F^N (u^M_{(i)} + c^M_{(i+1)}) = 0 \quad (1)$$

where F^N is the force component of the Nth variable, $u^M_{(i)}$ is the value of Mth variable after an iteration i and $c^M_{(i+1)}$ is the deviation of this solution from exact solution.
Using Taylor series expansion about $u^M_{(i)}$,

$$K^{NL}_{(i)}\, c^L_{(i+1)} = -\, F^N_{(i)} \quad (2)$$

In the above equation except the first two terms, rest are neglected if $c^M_{(i+1)}$ is very small,

where $K^{NL}_{(i)} = \dfrac{\partial F^N}{\partial u^L}(u^M_{(i)})$ is the system's tangent stiffness or Jacobian matrix and $F^N_{(i)} = F^N (u^M_{(i)})$.

The next approximation can be calculated by

$$\Delta u^M_{(i+1)} = \Delta u^M_{(i)} + c^M_{(i+1)} \quad (3)$$

Iterations are repeated in each increment until convergence is achieved.

2.2. *Boundary conditions*

Surface to surface interaction and multi-point tie constraint is provided between conduits and TL and between TL and support structure. The PTC is analysed at 0° and 90° collector angle under gravity load by providing the ENCASTRE Boundary Conditions (BCs) at appropriate places. The post processing of the results is done in MATLAB by following the elemental approach [6].

3. Results and Discussion

The basis of optimization is calculation of SDx value (in milliradians) which is also the statistical parameter characterizing the shape accuracy of the whole mirror surface and is calculated by the following formula which was mentioned by Lüpfert et al. [7].

$$SD_x = \sqrt{(\Sigma_{i,\,j=1}^{n}(sd_{x_{ij}}^2 \times \frac{a_{ij}}{A_{tot}}))} \tag{4}$$

where sd^2_{xij} are local slope deviation values, a_{ij}'s are the according element surface areas projected onto the aperture plane and A_{tot} is the aperture area. The mechanical properties used in relations of equation for law of mixtures have been mentioned in table 1 to get in-plane and transverse elastic constants for all the configurations [8],[9]. All the mechanical properties are calculated by using 50/50 V/V% of fibre/resin.

Table 1. Mechanical properties of fibres and resins used for modelling [8], [9].

	Elastic modulus (E) (in GPa)	Shear modulus (G) (in GPa)	Poisson's ratio (ν)	Density (ρ) (in Kg/m³)
Woven glass	72	29	0.25	2600
Woven Jute	25	7.24	0.38	1450
Unsaturated Polyester Resin	3.4	1.2	0.4	1110

3.1. *Optimization of size and direction of reinforced conduits*

The direction of reinforced conduits is one of the significant factors in calculation of trough surface deformations. From table 2, it is quite evident by the SDx values of all the three configurations that stiffness/weight ratio of the structure when the reinforcement is done in only X direction is higher than any other configuration under gravity loading. But this SDx value is very close to reinforced structure in both X and Y directions, so it is encouraged that latter configuration is used to counterbalance the wind loads from both yaw angle and pitch angle perspective because of its bi-directional strength. It can also be seen that reinforcement in only Y direction is least effective from structural point of view due to largest SDx value amongst the three configurations.

Table 2. Variation of SD_X values of all mirrors (in mrad) at fibre orientation of Δθ=45° of WFRPH trough model at 0° collector angle.

Reinforcement direction	Collector Angle		Mirror 1	Mirror 2	Mirror 3	Mirror 4	Mirror 5	Mirror 6	Mirror 7	Mirror 8	Average
X	0°	SDx	1.4064	1.1987	1.1917	1.4080	1.4006	1.1980	1.1961	1.4042	**1.3005**
	90°	SDx	0.5955	1.1032	1.0936	0.5880	0.5905	1.0026	1.0009	0.5895	**0.8205**
Y	0°	SDx	6.2265	3.2210	3.2216	6.2341	6.2230	3.2147	3.2081	6.2292	**4.7223**
	90°	SDx	1.5712	2.2172	2.2138	1.5740	1.5398	2.0495	2.0468	1.5458	**1.8448**
X,Y	0°	SDx	1.4957	1.2088	1.2031	1.4975	1.4946	1.2009	1.1978	1.4956	**1.3492**
	90°	SDx	0.6362	1.1063	1.0985	0.6279	0.6269	1.0131	1.0019	0.6244	**0.8419**

All the above three configurations are analysed by fixing the thickness of conduits as 0.75 mm and radius as 10 mm. The size of conduits is completely dependent upon the mesh size which is further dependent upon the convergence rate and the BCs. Also the laminate pattern is taken as $[0°_G/45°_G/-45°_J/90°_J]_s$ (0° is parallel to X direction) in all the three configurations since it is also the optimised configuration as discussed in the next section.

3.2. Optimization of TL

Finally the stacking sequence, its number and the orientation of fibres is optimised to achieve the best TL configuration and hence the least SDx value. Since in-plane mechanical properties of TL depends upon these factors, thus, to maintain homogeneity of its in-plane properties, TL is made symmetric and quasi-isotropic (orientation wise). As shown in the figure 3 and figure 4, the fibre orientation of Δθ=45° and Δθ=60° amongst the laminae (or layers) is studied separately by varying their stacking sequence and fixing the conduits in X and Y directions both. The number of layers are decided by taking the minimum thickness of laminae as per its practical viability and a total of 66 cases are studied in accordance to the symmetry of TL at Δθ=45° and Δθ=60°. For fibre orientation of Δθ=45°, the number of layers are taken as 8 where each layer is 0.5 mm thick and for fibre orientation of Δθ=60°, the number of layers are taken as 6 where each of the outer two layers from either sides are 0.5 mm thick. In general, at Δθ=45°, general notation of all the configurations can be $[h°_P/i°_P/j°_Q/k°_Q]_s$ where h/i/j/k exhibits the values 0/45/-45/90 respectively and P denotes G (or woven glass fibre and polyester lamina) whereas Q denotes J (or woven jute fibre and polyester lamina) and the sequence of both h/i/j/k and PQ can vary (variation of h/i/j/k is shown in table 3). Also interchanging only i, j in the same configuration (i.e. by keeping h, k and P, Q fixed) does not alter the results, so they are considered as one configuration which makes the total number of cases as 48 for Δθ=45°. Similarly at Δθ=60°, general notation of all the configurations can be $[a°_P/b°_P/c°_Q]_s$ where a/b/c exhibits the values 0/60/-60 respectively and PQ has the usual notations and total number of cases studied are 18 at Δθ=60°.

From figure 3 it is clear that $[0°_G/45°_G/-45°_J/90°_J]_s$ obtains the SDx value of 1.3492 mrad at Δθ=45° at 0° collector angle under gravity loading which is also lowest as compared to its rest of the configurations and the SDx value of 1.3503 mrad at $[0°_G/60°_J/-60°_J]_s$ configuration at Δθ=60° which is slightly higher than the above configuration. These SDx values are dependent upon interlaminate stresses and bending stresses produced as per theory of bending and classical laminate theory other than previously mentioned parameters like BCs, direction of reinforcement and collector's stiffness/weight ratio.

At Δθ=45°, from figure 4 it can be seen that at configurations K and L of TL in all the Glass and Jute sequences i.e. GGJJ, GJGJ, JGJG and JJGG, the surface deformation is maximum due to poor fiber sequence. According to Pagano and Pipes, interlaminar normal stresses (arises due to mismatch in moduli and Poisson's ratios between the layers) plays an important role in deciding the strength of the TL due to which SDx values are quite high in these configurations [10]. By similar reasoning, it can be said that configurations A and B have significantly lower SDx values due to their lesser tendency to delaminate. Similarly at Δθ=60°, the configuration A of TL in all the glass and jute sequences has shown the minimum SDx value amongst the rest of configurations but this minimum value is still higher than configuration A and B of Δθ=45° fibre orientation which makes it to be the optimum configuration.

Table 3. Different TL fibre orientations of WFRPH trough model.

Δθ=45°					
A 0/45/-45/90	B 0/45/-45/90	C 0/45/90/-45	D 90/45/0/-45	E 0/90/-45/45	F 90/0/-45/45
G 45/90/-45/0	H 45/0/-45/90	I 45/0/90/-45	J 45/90/0/-45	K 45/-45/90/0	L 45/-45/0/90

Δθ=60°		
A 0/60/-60	B 60/0/-60	C 60/-60/0

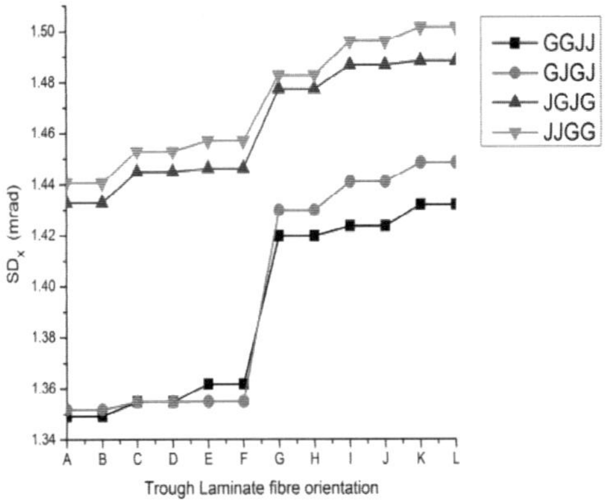

Figure 3. SDx values for different TL configurations when Δθ=45°.

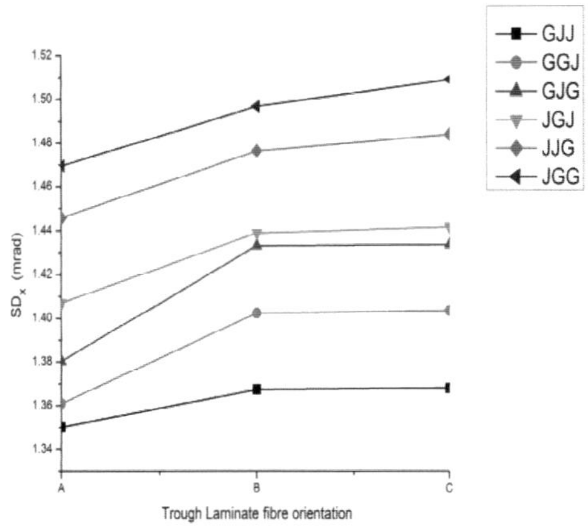

Figure 4. SDx values for different TL configurations when Δθ=60°.

In all the above configurations, WFRPH composite trough undergoes righted U and inverted U shaped deformations depending upon the BCs of the PTC, curvature effect of trough, collector orientation and its stiffness/weight ratio as shown in the figure 5 (a), (b).

(a)

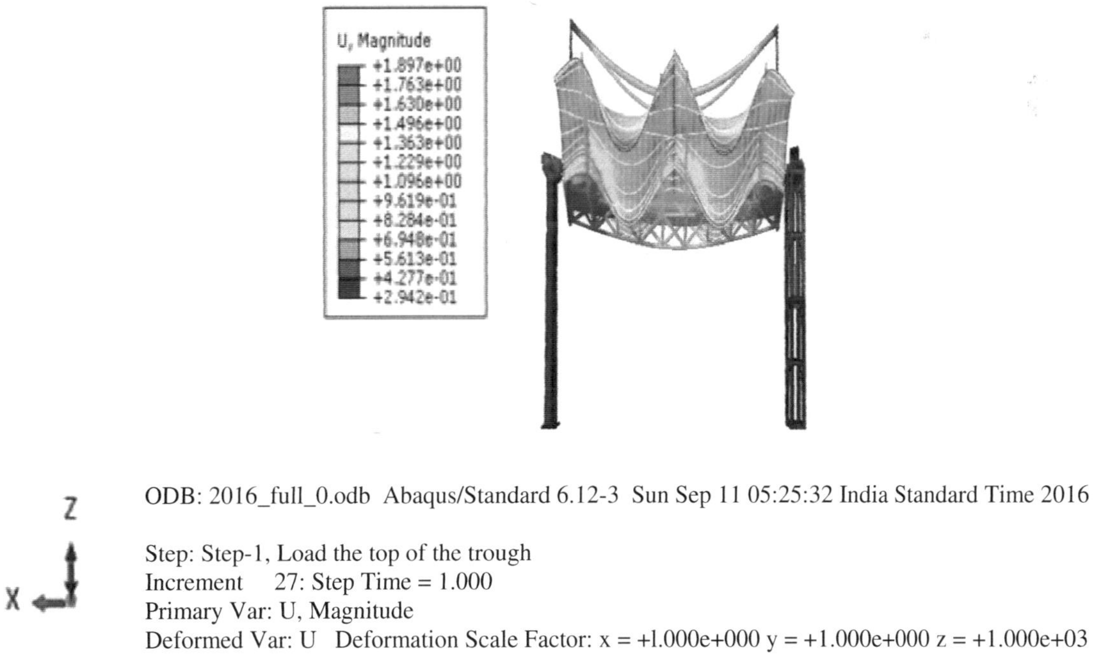

IOP Conf. Series: Materials Science and Engineering **222** (2017) 012016 doi:10.1088/1757-899X/222/1/012016

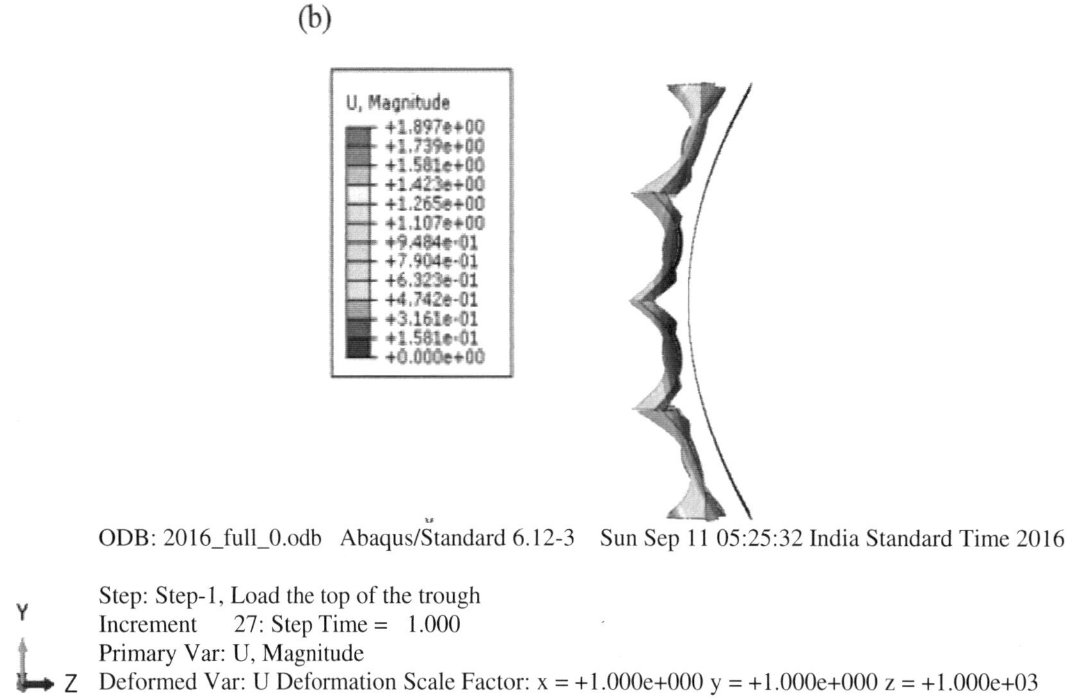

Figure 5. Deformation of WFRPH composite trough collector in 0° collector angle. (a) from its undeformed shape (superimposed) as shown in black in XZ plane (b) Trough deformation from its undeformed shape (superimposed) as shown in black in YZ plane. Color scale of deformation surface in mm, deformation scaling factor: 1000.

4. Conclusions

The deformation of cost effective solar PTC is studied using WFRPH composite to improve its optical efficiency. To ensure high stiffness/weight ratio and thus lower SDx values, its optimization is done by varying different trough parameters viz. direction and size of reinforced conduits, stacking number and sequence of hybrid TL at fibre orientation of Δθ=45° and Δθ=60° amongst the laminae at 0° collector angle. The analysis shows that reinforcement in the X direction is the most effective amongst the rest of the configurations under gravity load but it is recommended that reinforcement is done in both the directions to scale down the effect of wind loads. Furthermore it can also be seen that laminate pattern of $[0°_G/45°_G/-45°_J/90°_J]_s$ undergoes minimum surface deformation when reinforced in X and Y directions both due to the presence of reduced interlaminar stresses in this configuration. Though SDx values are strongly dependent upon the parameters like collector's stiffness/weight ratio, curvature of trough, and the BCs but the results proves that arguably WFRPH PTC is the promising substitute of conventional trough collector and hence can be easily used for compensating the costs required to fabricate the mould in batch production of parabolic trough reflectors.

References

[1] Kalogirou, S., Eleftheriou, P., Lloyd, S., Ward, J (1994). Low cost high accuracy parabolic troughs construction and evaluation. *Renewable Energy, 5 (1-4),* 384-386.

[2] Valan Arasu, A., Sornakumar, T (2007). Design, manufacture and testing of fiberglass reinforce parabola trough for parabolic trough solar collectors. *Solar Energy, 81 (10),* 1273-1279.

[3] Coccia, G., Di Nicola, G., Sotte, M. (2014). Design, manufacture, and test of a prototype for a parabolic trough collector for industrial process heat. *Renewable Energy, 74,* 727-736.

IOP Conf. Series: Materials Science and Engineering **222** (2017) 012016 doi:10.1088/1757-899X/222/1/012016

[4] Abel Ezimokhai, Alex Chong, Jason Matthews (2015). Modelling and Assessment of Repairs Conducted on Composite Materials. *International J. Mater. Science and Engg., No. 3,* 252-266.

[5] Hibbit, Karlsson, Sorensen (2010). Abaqus Analysis User's Manual, vol 2-Abaqus 6.10.

[6] Meiser, S., et. al. (2013). Finite Element Modeling of Parabolic Trough Mirror Shape in Different Mirror Angles. *J. Solar Energy Engg.*; 135 031006.

[7] Lupfert, E., Ulmer, S. (2009). Solar Trough Mirror Space Specifications. *Proceedings of 15th SolarPACES Conference, Berlin (Germany).*

[8] Akkerman, Remko (2006). *Laminate mechanics for balanced woven fabrics. Composites Part B:Engineering, 37 (2-3),* 108-116.

[9] S. Belaid, S.F. Chabira, P. Balland, M. Sebaa, (2015). Thermal aging effect on the mechanical properties of polyester fiberglass composites. *J. Mater. Environ. Sci. 6 (10),* 2795-2803.

[10] N.J. Pagano, R. Byron Pipes (1971). The Influence of Stacking Sequence on Laminate Strength. *Journal of Composite Materials,* pp. 50-57.

IOP Conf. Series: Materials Science and Engineering **222** (2017) 012017 doi:10.1088/1757-899X/222/1/012017

Reliability evaluation of microgrid considering incentive-based demand response

Ting-Cheng Huang, Yong-Jun Zhang

School of Electricity Power, South China University of Technology, Wushan Road, Guangzhou, China

E-mail: 339536312@qq.com

Abstract. Incentive-based demand response (IBDR) can guide customers to adjust their behaviour of electricity and curtail load actively. Meanwhile, distributed generation (DG) and energy storage system (ESS) can provide time for the implementation of IBDR. The paper focus on the reliability evaluation of microgrid considering IBDR. Firstly, the mechanism of IBDR and its impact on power supply reliability are analysed. Secondly, the IBDR dispatch model considering customer's comprehensive assessment and the customer response model are developed. Thirdly, the reliability evaluation method considering IBDR based on Monte Carlo simulation is proposed. Finally, the validity of the above models and method is studied through numerical tests on modified RBTS Bus6 test system. Simulation results demonstrated that IBDR can improve the reliability of microgrid.

1. Introduction

Microgrid is a micro power distribution system that integrates DG, ESS, load, converter and protection device[1]. Microgrid can reduce the adverse effects of intermittent DG, improving power supply reliability and power quality. The reliability of microgrid is an important research content.

Demand response(DR) refers to the behavior that users adjust their power consumption actively according to the price signal or incentive mechanism[2]. According to the different response mechanisms, DR can be divided into time-based rates DR and IBDR[3]. IBDR refers to users get payments or preferential prices form reducing power usage during periods of system need or stress.

At present, the research on DR mainly focused on the mechanism analysis of DR[3], DR modeling for large-scale clean energy consumption[4], the benefit evaluation and support technology of DR[5] and so on. IBDR usually takes some time to respond and cannot reduce the load immediately to maintain a balance between power supply and demand in the event of a grid failure. Therefore, IBDR is usually not considered in the reliability assessment. In the microgrid, DG and ESS can provide response time for the implementation of IBDR to reduce load when microgrid switches to island mode. Loads of the microgrid are powered by DG and ESS before the load is curtailed.

In this paper, a reliability evaluation method of microgrid considering the IBDR is proposed. The basic idea is as follows: a) The dispatch model and response model on the IBDR are developed; b) Monte Carlo method is used to simulate the operating state of grid elements; c) Solve the dispatch model to get the amount of load curtailed by customers when microgrid is running on an island state; d) Solve the response model to get the actual amount of load reduced by each customer, and then determine the operating state of load points; e) Evaluate the reliability of microgrid.

2. Mechanism of IBDR

The IBDR programs studied in the paper includes interruptible load (IL) and emergency DR (EDR).

Customers in IL programs receive a rate discount or bill credit in exchange for agreeing to reduce load during system contingencies[3]. Customers can be penalized if they do not curtail. IL program has become an important tool for peak load shifting and handing of emergencies.

EDR program is implemented in a manner similar to that of IL program, where an incentive payment is set by the utility company and customer can obtain the compensation by reducing the load in the event of a reliability incident[6]. EDR is generally not used for peak load shifting. Customers can also choose to forgo the incentive payment and not curtail without penalty.

3. IBDR dispatch model

3.1. Comprehensive evaluation of customer in IBDR program

In this paper, the time to complete the response (M1), the response rate (M2) and the maximum load reduction (M3) are taken into account for the comprehensive evaluation of IBDR customer. M1 refers to the time from the moment that customer receives the response signal to the moment that load reduction is completed. M2 refers to the proportion of the number of times that customer actually curtailed the load in the total number of times that signal was received. M3 refers to the maximum amount of load that customer can curtail. The calculation method of M3 is shown as Equation (1).

$$M_3 = P_{j,\max}\lambda_{j,\max} \tag{1}$$

where $P_{j,\max}$ is the jth customer's peak load; $\lambda_{j,\max}$ is the jth customer's maximum load reduction ratio specified in the agreement.

The entropy method is used to determine the weight of the index, and the calculation steps are as follows: a) Standardize evaluation index; b) Calculate the entropy of the index; c) Calculate the entropy weight of the index. Then the jth customer's comprehensive evaluation value is as following:

$$W_j = \sum_{f=1}^{3} \omega_f b_{jf} \tag{2}$$

where ω_f is the entropy weight of the fth index; b_{jf} is the standardized result of the jth customer's fth index, f=1,2,3, corresponding to M_1, M_2 and M_3 respectively.

3.2. Principle of customer load reduction

After the microgrid enters the island mode, the internal load of the microgrid is supplied by DG and ESS. At this time, the unbalanced power is mainly borne by ESS and the output stable DG, such as diesel engine and micro gas turbine. ESS cannot be a long time power supply due to limited capacity. Therefore, it's necessary to allocate the output power of ESS to the customers in DR programs in advance. The allocation method is shown as Equation (3).

$$\Delta P = P_{\mathrm{ST}} = \sum_{d=1}^{D} P_{\mathrm{DG}d} - \sum_{d=1}^{D} P_{\mathrm{L}d} - P_{\mathrm{loss}} \tag{3}$$

where ΔP is unbalanced power assigned to the customer; P_{ST} is the output power of ESS; D is the total number of nodes; $P_{\mathrm{DG}d}$ is the DG output power of dth node; $P_{\mathrm{L}d}$ is the load power of dth node; P_{loss} is loss of microgrid. As the ESS output power changes with time, ΔP can be calculated by Equation (4).

$$\Delta P = \overline{P_{\mathrm{ST}}} = \int_{t_1}^{t_2} P_{\mathrm{ST}}(t)dt / (t_2 - t_1) \tag{4}$$

where $\overline{P_{\mathrm{ST}}}$ is the average output of ESS; $P_{\mathrm{ST}}(t)$ is a function of ESS output over time; t_1 and t_2 are the starting and ending time for averaging ESS output.

3.3. IBDR dispatch model

The dispatch department assigns unbalanced power to customers during system contingencies. In order to improve the supply reliability and reduce the risk of customer's default, the objective function of dispatch model is to maximum the average evaluation value of customers that are dispatched. The dispatch model is shown as following:

$$\begin{aligned} &\max \overline{W} = \sum_{i=1}^{n}(W_i \Delta P_i) / \sum_{i=1}^{n} \Delta P_i \\ \text{s.t.} \quad &\sum_{i=1}^{n} \Delta P_i = \omega_{\mathrm{a}} \Delta P \\ &\max(M_{1,i}) \le T_{\mathrm{S}},\ i \in \mathbf{N} \\ &\forall i \in \mathbf{N}, \Delta P_i / P_i \le \lambda_{i,\max} \end{aligned} \tag{5}$$

where n is the number of IBDR customers; $\overline{W}$ is the average evaluation value of customers; ΔP_i is load reduction of ith customer; ω_{a} is the margin of the load reduction and $\omega_{\mathrm{a}} \ge 1$; $M_{1,i}$ is ith customer's M_1; $\mathbf{N}$ is the set of customers; T_{S} is the remaining available time of ESS; P_i is ith customer's load before the response.

T_{S} mainly depends on the capacity and output of ESS and is calculated by Equation (6).

$$T_{\mathrm{S}} = \left(Q_{\mathrm{re}} - Q_{\min}\right) / \overline{P_{\mathrm{ST}}} \tag{6}$$

where Q_{re} is the remaining capacity of ESS; $Q_{\min}$ is the minimum capacity limit for ESS.

The duration of customer's response (t_{d}) can be calculated as following:

$$t_{\mathrm{d},i} = \omega_{\mathrm{b}} T_{\mathrm{d}} \tag{7}$$

where ω_{b} is response time margin coefficient and $\omega_{\mathrm{b}} \ge 1$; T_{d} is the expected repair time.

4. Customer response model

The response behavior of customer depends on whether customer can profit from the response. The paper develops a unified response model for customers in IL and EDR programs, which is based on the demand side goal optimization method and takes into account the uncertainty of DR. The customer response model is shown as following:

$$\begin{aligned} &\max L = R - C_1 - C_2 - F \\ &\begin{cases} R = \begin{cases} \Delta P_{\mathrm{n}} E t_{\mathrm{d}} & \Delta P_{\mathrm{a}} \ge \Delta P_{\mathrm{n}} \\ \Delta P_{\mathrm{a}} E t_{\mathrm{d}} & \Delta P_{\mathrm{a}} < \Delta P_{\mathrm{n}} \end{cases} \\ C_1 = (K_1 \Delta P_{\mathrm{a}}^2 + K_2 \Delta P_{\mathrm{a}} - K_2 \Delta P_{\mathrm{a}} u) t_{\mathrm{d}} \\ C_2 = \int_{t_0}^{t_0 + t_{\mathrm{d}}} \alpha p (P(t) - \Delta P_{\mathrm{a}}) dt \\ F = \begin{cases} 0 & \Delta P_{\mathrm{a}} \ge \Delta P_{\mathrm{n}} \\ (\Delta P_{\mathrm{n}} - \Delta P_{\mathrm{a}}) p_{\mathrm{f}} t_{\mathrm{d}} & \Delta P_{\mathrm{a}} < \Delta P_{\mathrm{n}} \end{cases} \end{cases} \\ \text{s.t.} \quad &0 \le \Delta P_{\mathrm{a}} \le P(t_0) \\ &\alpha = \begin{cases} \alpha & \Delta P_{\mathrm{a}} \ge \Delta P_{\mathrm{n}} \\ 1 & \Delta P_{\mathrm{a}} < \Delta P_{\mathrm{n}} \end{cases} \\ &t_{\mathrm{s}} \le t_{\mathrm{sn}} \end{aligned} \tag{8}$$

where L is total economic benefits; R is the response gain; C_1 is the loss of power outage; C_2 is the electricity cost during the response period; F is the unresponsive penalty; ΔP_{n} is the amount of load reduction required by the dispatch department; ΔP_{a} is the actual load reduction; E is the compensation for the unit load reduction; K_1 and K_2 are constant coefficients; u is a random value in the (0,1) interval; t_0 is the time to start the response; α is the discount rate; p is the price; $P(t)$ is a function of the load over time; p_f is penalty factor; t_{s} is the time to complete the response; t_{sn} is the time to complete the response of the agreement.

5. Reliability evaluation method considering IBDR

Based on the Monte Carlo simulation, this paper presents a microgrid reliability evaluation method considering IBDR. The concrete steps are as follows:

1) Initialize the data, set the simulation years; 2) Generate the time to failure (TTF) for each grid component; 3) Select the component whose TTF is least as a faulty component, set its number to m, then the system's normal working time $T_{TF\mathrm{s}} = T_{TF\mathrm{m}}$; 4) Generate the time to repair (TTR) for

component *m*; 5) Analyze the impact of component *m* failure on the load and divide the load into 5 categories. Class A: load not affected by the failure; Class B: load whose power supply can be restored after the fault has been isolated; Class C: load whose power supply can be restored through transfer the power; Class D: non-transferable load; Class E: microgrid load. 6) Class A load does not fail; The B, C and D class loads power outage time are fault isolation time, load transfer time and fault repair time; If microgrid does not enter the island mode, E class load does not fail; If not, proceed to the following steps:

a) Let L_1 be the set of loads that are cut and let L_2 be the set of loads that are not cut. Get the maximum output P_M and remaining capacity Q_{re} of ESS. b) Calculate the demand power $P_i(t)$ of each load point and total power in the L_2 set, DG output power $P_{DG}(t)$ and loss in t to $t+\Delta t(\Delta t$=15min) time period. c) If $P_{DG}(t)>P_L(t)+P_{loss}$, go to step i. d) If $P_{DG}(t)+P_M<P_L(t)+P_{loss}$, go to step j. e) If $T_S>\Delta t$ +max($t_{sn,i}$), go to step i. f) Calculate the maximum load reduction $\Delta P_{max}=\sum_{i=1}^{n}P_i(t)\lambda_{i,max}$ .If $P_{DG}(t)>P_L(t)+P_{loss}-\Delta P_{max}$, go to step h. g) If $T_S<\Delta t$, go to step j. If not, go to step k. h) Solve the dispatch model and send the response signal to the customer. i) Solve the response model. If $\sum\Delta P_{a,i}\geq\Delta P$, $Q_{re}=Q_{re}-\overline{P_{ST}}\max(t_{sn,i})$, $t=t+\max(t_{sn,i})$, and go to step l. If not, go to step g. j) Cut the load according to the importance and the position of the load and save the set L_1 and L_2. k) $Q_{re}=Q_{re}-\overline{P_{ST}}\max(t_{sn,i})$. l) If t is equal to the repair time, calculate the reliability index of the loads in the set L_1 and end this cycle. If not, $t=t+\Delta t$ and go to step b.

7) If the simulation time does not reach the set time, go to step 2. 8) Calculate the reliability index of the load point and system.

6. Analysis of example

The modified RBTS Bus6 F4 feeder system is used for reliability evaluation and shown as figure 1.

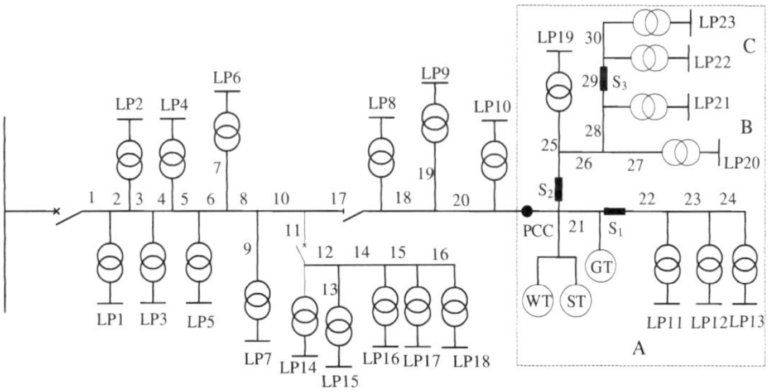

Figure 1. Modified RBTS Bus6 F4 feeder system

WT is the wind turbine and its maximum output is 1MW. GT is the gas turbine and its rated output is 0.6MW. ST is the energy storage whose capacity is 1MW·h and output is 0.5MW. PCC is the common connection point between microgrid and distribution network. S_1, S_2 and S_3 are intelligent switches. The reliability parameters of the components are given in the reference[7]. The load data are given in the reference[8].

The following 3 cases are set up for reliability assessment and analysis. Case 1: no microgrid and no IBDR; Case 2: there is a microgrid but no IBDR; Case 3: there are microgrid and IBDR. Select LP 1, 9, 11, 20 and 22 as typical loads. Their annual failure rate and annual average interruption duration are shown in figure 2. The reliability indices of microgrid are shown in table 1.

Table 1 Reliability indices of microgrid

Case	SAIFI	SAIDI	ASAI	EENS
2	1.3689	9.6638	99.8897	17.233
3	0.8956	7.3511	99.9161	13.588

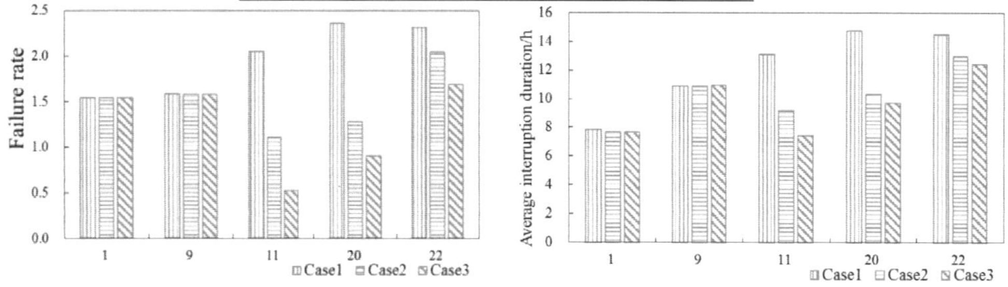

Figure 2. Reliability index of typical load points

In table 1, SAIFI is the system average interruption frequency index, SAIDI is the system average interruption duration index, ASAI is the average service availability index and EENS is the expected energy not supplied.

As can be seen from the simulation:

1) The implementation of IBDR can improve the reliability level of microgrid.

2) In contrast to Case 2 and Case 3, it can be seen that the implementation of IBDR can further improve the reliability of the microgrid under the same conditions of DG and ESS.

3) Compared with the reliability indices of the typical load points, it can be seen that the level of reliability improved by IBDR is related to the importance of the load. The more important the load is, the higher the degree of reliability improved by IBDR is.

7. Conclusions

Based on the characteristics of microgrid and IBDR, this paper applies the IBDR to the reliability evaluation of microgrid, proposes the dispatch model and response model of IBDR and evaluate the reliability of the modified RBTS Bus6 test system based on Monte Carlo simulation. The results show that the implementation of IBDR can improve the reliability of microgrid. However, IBDR is only one type of DR. In the next study, we need to study the optimal allocation and coordination control of DG, ESS and demand side resources to improve the reliability in combination with the time-based rates DR.

References

[1] Hatziargyriou, Nikos, et al 2007 Microgrids *J. IEEE power and energy magazine* **5** 78-94.

[2] Kirschen D S 2003 Demand-side view of electricity markets *J. IEEE Transactions on Power Systems* **18** 520-27.

[3] Zhong H, Xie L and Xia Q 2013 Coupon incentive-based demand response: Theory and case study *J. IEEE Transactions on Power Systems* **28** 1266-76.

[4] Siano P 2014 *J. Renewable and Sustainable Energy Reviews* **30** 461-78.

[5] Yan H, Li B, C S, et al 2015 *J. Journal of Modern Power Systems and Clean Energy* **3** 72-81.

[6] Molina-Garcia A, Bouffard F and Kirschen D S 2011 *J. IEEE Transactions on Power Systems* **26** 411-19.

[7] Allan R N, Billinton R, Sjarief I, et al 1991 *J. IEEE Transactions on Power Systems* **6** 813-20.

[8] Billinton R and Jonnavithula S 1996 *J. IEEE Transactions on Power Systems* **11** 1670-76.

Acknowledgments

This work is supported by National Natural Science Foundation of Chian (No. 51377060).

IOP Conf. Series: Materials Science and Engineering **222** (2017) 012018 doi:10.1088/1757-899X/222/1/012018

Real-time pricing strategy of micro-grid energy centre considering price-based demand response

Zhiheng Xu, Yongjun Zhang, Gan Wang

South China University of Technology, Wushan road, Tianhe District, Guangzhou, Guangdong province, China

Email: 1779496063@qq.com

Abstract. With the development of energy conversion technology such as power to gas (P2G), fuel cell and so on, the coupling between energy sources becomes more and more closely. Centralized dispatch among electricity, natural gas and heat will become a trend. With the goal of maximizing the system revenue, this paper establishes the model of micro-grid energy centre based on energy hub. According to the proposed model, the real-time pricing strategy taking into account price-based demand response of load is developed. And the influence of real-time pricing strategy on the peak load shifting is discussed. In addition, the impact of wind power predicted inaccuracy on real-time pricing strategy is analysed.

1. Introduction

With the development of economy and demand, the energy dilemma and environmental pollution becomes more and more severe. As a new energy consumption pattern, Energy Internet has received more attention which can improve energy efficiency [1,2]. Nowadays, electricity, natural gas and other energy network is independent in plan and operation. As the development of energy converter, such as power to gas (P2G) and fuel cell, the coupling among electricity, natural gas and heat network become closer and closer [3,4]. Based on it, micro-grid energy centre combines the electricity, natural gas and heat and realizes centralized dispatch will become a trend [5,6].

There are several proposals for energy optimal operation based on energy hub. In [7], optimal energy flow model in electricity-gas integrated energy system was proposed. In [8], the impact of P2G in electricity-gas integrated energy system was considered. These proposals do not consider the impact of load price-based demand respond on real-time pricing strategy and system revenue.

In the other hand, access to renewable energy will have an impact on the operation of micro-grid energy centre. In this paper, optimal energy dispatch model of micro-grid energy centre is established. Based on this model, the real-time pricing strategy considering price-based demand response of load is developed. Furthermore, the impact of wind power predicted inaccuracy on real-time pricing strategy is analyzed.

2. Model description

2.1. Static-state model of micro-grid energy centre

The structure of micro-grid energy centre is shown in figure 1. Typical elements composing a micro-grid energy centre are renewable energy, energy converters and energy storages. The $\boldsymbol{P}$, $\boldsymbol{L}$, $\boldsymbol{R}$, $\boldsymbol{Q}$, $\boldsymbol{S}$ and $\boldsymbol{\eta}$ are represent the input energy flow form network side, output energy flow, renewable energy sources, charge and discharge power of energy storages, input energy flow of energy converters and energy convert efficiency respectively. The total number of variables $\boldsymbol{P}$, $\boldsymbol{L}$, $\boldsymbol{S}$, $\boldsymbol{R}$, $\boldsymbol{Q}$ are defined as M, N, I, K and H respectively. The static-state balanced equations are shown as formula (1).

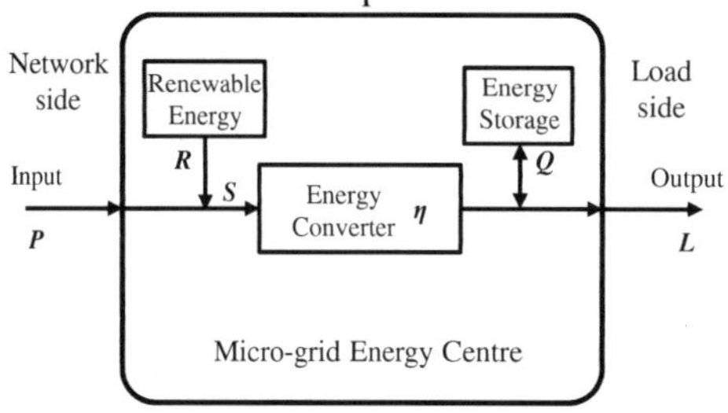

Figure 1. Structure of Micro-grid Energy Centre.

$$\begin{bmatrix} \boldsymbol{Z}_{\mathrm{A}} \\ \boldsymbol{Z}_{\mathrm{B}} \end{bmatrix} \begin{bmatrix} \boldsymbol{P} \\ \boldsymbol{S} \\ \boldsymbol{R} \\ \boldsymbol{Q} \end{bmatrix} = \begin{bmatrix} \mathbf{1} & \boldsymbol{Z}_{\mathrm{AS}} & \boldsymbol{Z}_{\mathrm{AR}} & \mathbf{0} \\ \mathbf{0} & \boldsymbol{Z}_{\mathrm{BS}} & \mathbf{0} & \boldsymbol{Z}_{\mathrm{BQ}} \end{bmatrix} \begin{bmatrix} \boldsymbol{P} \\ \boldsymbol{S} \\ \boldsymbol{R} \\ \boldsymbol{Q} \end{bmatrix} = \begin{bmatrix} \mathbf{0} \\ \boldsymbol{L} \end{bmatrix} \tag{1}$$

In (1), $\boldsymbol{Z}_{\mathrm{A}}$ and $\boldsymbol{Z}_{\mathrm{B}}$ are defined as source incidence matrices and load incidence matrix. **1** is an $M \times M$ unit matrix. Partitioned matrices of $\boldsymbol{Z}_{\mathrm{A}}$ and $\boldsymbol{Z}_{\mathrm{B}}$ include $\boldsymbol{Z}_{\mathrm{AS}}=(z_{mi})_{M\times I}$、 $\boldsymbol{Z}_{\mathrm{AR}}=(z_{mk})_{M\times K}$、 $\boldsymbol{Z}_{\mathrm{BS}}=(z_{ni})_{N\times I}$ and $\boldsymbol{Z}_{\mathrm{BQ}}=(z_{nh})_{N\times H}$. These partitioned matrices also can be defined as formulas (2).

$$\begin{cases} z_{mi} = \begin{cases} -1 & m \text{ connects with } i \\ 0 & \text{else} \end{cases} \\ z_{mk} = \begin{cases} 1 & m \text{ connects with } k \\ 0 & \text{else} \end{cases} \end{cases} \quad \begin{cases} z_{ni} = \begin{cases} \eta_{ni} & n \text{ connects with } i \\ 0 & \text{else} \end{cases} \\ z_{nh} = \begin{cases} -1 & n \text{ connects with } h \\ 0 & \text{else} \end{cases} \end{cases} \tag{2}$$

2.2. Model of load considering demand respond

Energy has pricing demand elasticity when it participates in the transaction as a commodity. The pricing demand elasticity is a predetermined parameter which shows energy demand changing with different price. The model of demand can be described as follows.

$$L_t = E_d \rho_t + \beta \tag{3}$$

$$\underline{\rho} \le \rho_t \le \overline{\rho} \tag{4}$$

In formula (4) and (5), E_d is the pricing demand elasticity. The β is a predetermined parameter. Subscript t is the counting variable for time period. The ρ_t represents energy sale price. The $\overline{\rho}$ and $\underline{\rho}$ correspond to upper/lower limits of the ρ_t.

3. Day ahead optimal dispatch of micro-grid energy centre

3.1. Objective function

$$F = \max \sum_{t=1}^{T} \left[\sum_{n=1}^{N} \rho_{n,t} L_{n,t} - \sum_{m=1}^{M} \theta_{m,t} P_{m,t} \right] \tag{5}$$

In equation (6), T is the total number time periods for scheduling. Subscripts, such as m and n are the counting variable for energy input and output. The $\rho_{n,t}$ and $\theta_{m,t}$ represent energy price of output and input respectively.

3.2. Constraints

Operating constraints of micro-grid energy centre include formula (1), (4), constraints of exchange power with networks side and operating constraints of energy converters, renewable energy resources and energy storages. The constraints can be represented as follows.

$$\begin{cases} 0 \le P_m \le P_m^{\max} \\ 0 \le R_{k,t} \le R_{k,t}^{*} \\ S_j^{\min} \le S_{j,t} \le S_j^{\max} \\ -S_j^{\text{down}} \le S_{j,t} - S_{j,t-1} \le S_j^{\text{up}} \\ E_h(t+1) = E_h(t) + Q_h^{\text{ch}}(t) - Q_h^{\text{dis}}(t) \end{cases} \quad \begin{cases} E_h(t)^{\min} \le E_h(t) \le E_h(t)^{\max} \\ E_h(0) = E_h(T) \\ 0 \le Q_h^{\text{ch}}(t) \le \varsigma_h^{\text{ch}}(t) Q_h^{\text{ch,max}}(t) \\ 0 \le Q_h^{\text{dis}}(t) \le \varsigma_h^{\text{dis}}(t) Q_h^{\text{dis,max}}(t) \\ \varsigma_h^{\text{ch}}(t) + \varsigma_h^{\text{dis}}(t) \le 1 \end{cases} \tag{6}$$

In formula (6), $R_{k,t}^{*}$ is renewable energy power forecast. S_j represents capacity of converters. Subscripts, such as max and min correspond to upper and lower limits of the variable. S^{up} and S^{down} are the upward and downward ramping capacity of converters. $Q_{\text{ch}h}$ and $Q_{\text{dis}h}$ presents charge and discharge power of energy storages respectively. E is energy of energy storage. ς_h^{ch} (t) and ς_h^{dis} (t) are the binary variables.

4. Case study

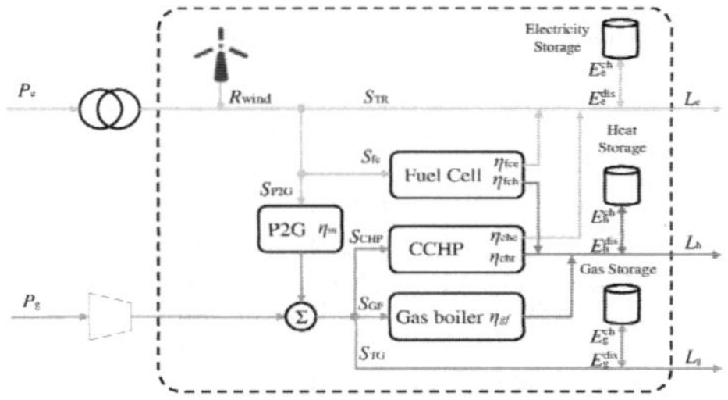

Figure 2. The Model of Micro-grid Energy Centre.

Figure 2 shows the model of micro-grid energy centre. Energy converters include P2G facilities, combined cooling heating and power facilities (CCHP), gas boilers and fuel cells. Parameters of load demand curves are shown as table 1.

Table 1. Parameters of Load Demand Curves.

Type of load	Basic price ($/MWh)	Demand elasticity E_d (MWh2/$)	Lower limits of price ($/MWh)	Upper limits of price ($/MWh)
electric	90	0.11	75	90
		0.06	90	120
heat	50	0.03	40	75
natural gas	80	0.06	70	100

This paper assumes that demand in basic price is called original load. The wind power forecast and original load are shown as figure 3.The prices of purchasing electricity and gas are shown as figure 4.

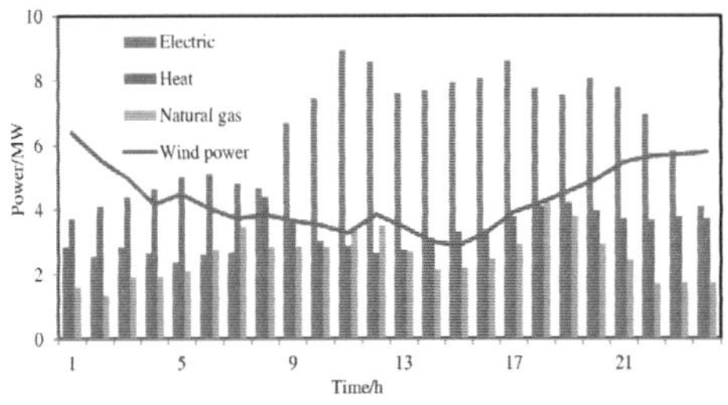

Figure 3. Wind Power Forecast and Original Load.

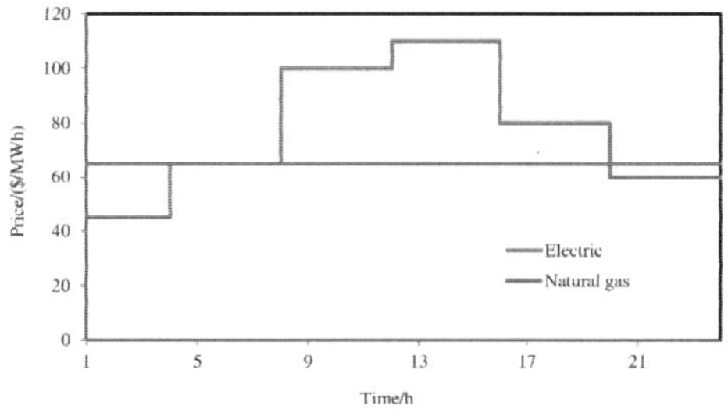

Figure 4. Prices of Purchasing Electricity and Gas.

4.1. Result analysis of optimal dispatch

According to the above data, the real-time pricing strategy, optimized load and energy purchase situation are shown as figure 5 and figure 6.

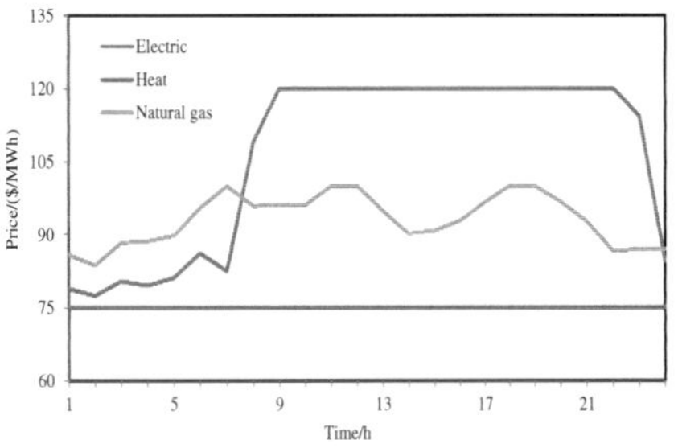

Figure 5. Real-time Pricing StrategyFigure.

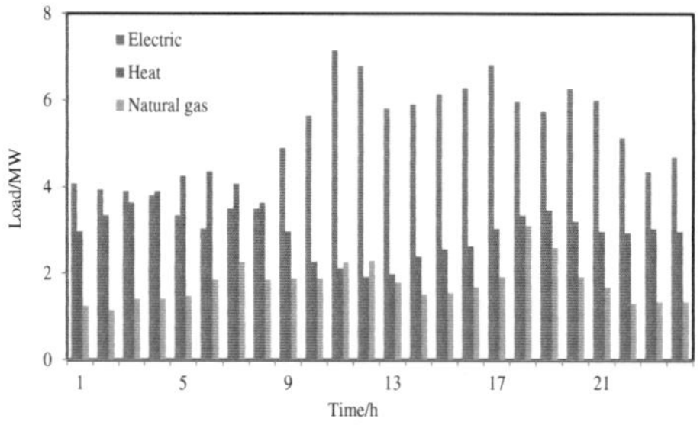

Figure 6. Energy load in Optimal Dispatch.

The 1 o'clock to 7 o'clock and the 21 o'clock to 24 o'clock are the electric trough. At the same time, the wind power is at its peak. In order to achieve the objective of the maximal profit, the electric sale price remains at around 80 $/MWh which has declined compared to the basic price. The 8 o'clock to 20 o'clock are the trough of wind power and peak of electricity. At this time, the purchase price of electricity is at its peak. Thus, the electric sale price almost is set in the upper limits of price to reduce the electric load. The purchase price of natural gas is stable. So the natural gas sale price is positively related to gas demand. The cost of heat is high because heat load can only be provided by electricity and natural gas after conversion. So the heat sale price always is set in the upper limits of price.

Comparison of the variances between the optimized load and the original load is shown as table 2.

Table 2. Comparison of Variances between the Optimized load and the Original load.

Type of load	Variances of original load	Variances of optimized load
electric	2.39	1.22
heat	0.67	0.67
natural gas	0.75	0.46

The variances of the optimized electric and gas load have declined compared to the original. The variance of heat load has not changed because the heat sale price is set in the upper limits of price. Therefore, implementation of real-time price has a significant effect on peak load shifting.

4.2. The impact of predicted inaccuracy on pricing strategy

This paper assumes the wind power in 4.1 is the real wind power and the corresponding pricing strategy is the standard strategy. The impact of ±15% and ±30% wind power predicted inaccuracy on system revenue and pricing strategy are as follows.

Table 3. System Revenue in Different Predicted Inaccuracy.

Predicted Inaccuracy	0	+15%	-15%	+30%	-30%
Revenue /$	12268.85	13208.50	11268.85	14066.04	10215.74
Revenue Inaccuracy	0	7.66%	-8.15%	14.6%	-16.7%

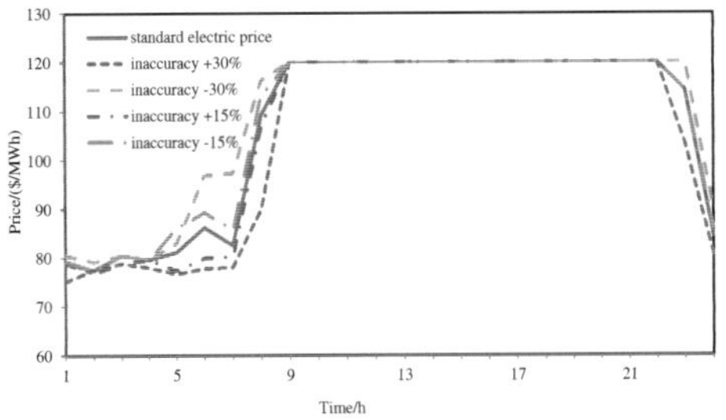

Figure 7. Electric Pricing Strategy in Different Predicted Inaccuracy.

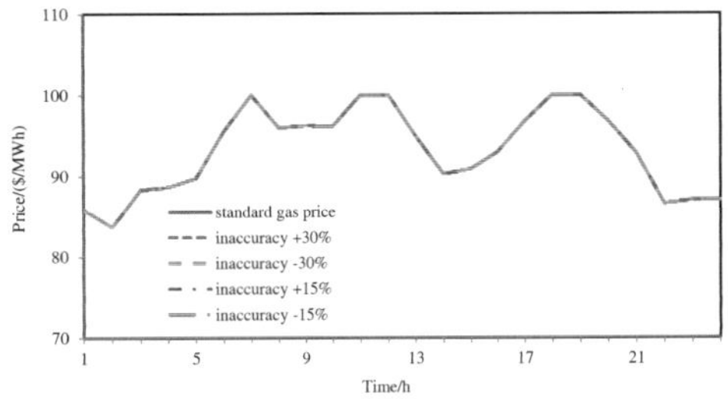

Figure 8. Natural gas Pricing Strategy in Different Predicted Inaccuracy.

Table 3, figure 7 and figure 8 show the revenues and the pricing strategies of electricity and gas under different inaccuracies. Gas load is mainly provided by gas sources because of low capacity of P2G facility. So predicted inaccuracy of wind power has no effect on gas price. In electric trough, electric load is mainly provided by wind power. Thus, predicted inaccuracy of wind power has a significant effect on the electric price in electric trough. Furthermore, the result also shows that the higher the inaccuracy, the greater the price fluctuation.

5. Conclusion

Based on model of micro-grid energy centre, this paper establishes optimal scheduling model considering price-based demand response, which aims at the maximum revenue. In addition, the proposed model has been tested to make the optimal real-time pricing strategy. And extensive simulation studies have indicated the following:

(1) Implementation of real-time price based on micro-grid energy centre is beneficial to peak load shifting.

(2) It is necessary to improve the wind power forecasting method and prediction accuracy of wind power in order to make real-time electric pricing strategy exactly.

(3) Uncertainties of wind power and energy load have a significant effect on real-time pricing strategy. Thus, the real-time price strategy considering these uncertainties needs further study.

References

[1] Li R, Chen L, Yuan T and Li C L 2016 *J. Mod. Power Syst. Clean Energy.* **4** 566
[2] Krause T, Andersson G, Frohlich K and Vaccaro A 2011 *Proc. of the IEEE* **99** 15
[3] Clegg S and Mancarella P 2016 *IET Gen., Trans. & Dist.* **10** 566
[4] Götz M, Lefebvre J, Mörs F, Koch M D, Graf F, Bajohr S 2016 *IET Gen., Trans. & Dist.* **85** 1371
[5] Vaccaro A, Pisani C and Zobaa A 2015 *IET Gen., Trans. & Dist.* **9** 1544
[6] Moeini-Aghtaie M, Farzin H, Fotuhi-Firuzabad M and Amrollahi R 2017 *IEEE Trans. on Power Sys.* **32** 368
[7] Chen S, Wei Z, Sun G, Cheung K, and Wang D 2016 *IEEE Trans. on Sust. Energy* **9** 1
[8] Clegg S and Mancarella P 2015 *IEEE Trans. on Sust. Energy* **6** 1234

Acknowledgments
This work is supported by National Natural Science Foundation of Chian (No. 51377060).

IOP Conf. Series: Materials Science and Engineering **222** (2017) 012019 doi:10.1088/1757-899X/222/1/012019

A Diarylethene Derived Chemosensor for Colorimetric Sensing of Cu^{2+}

Shouzhi Pu,[a] Qi Sun, Chunhong Zheng[a]

Jiangxi Key Laboratory of Organic Chemistry, Jiangxi Science and Technology Normal University, 30013 Nanchang Jiangxi PR China

E-mail: [a] pushouzhi@tsinghua.org.cn (S. Pu), ezirobot@163.com (C. Zheng)

Abstract. A diarylethene bearing a 8-hydroxyquinoline-linked salicylhydrazide Schift base unit has been synthesized. In CH_3CN, the conpound displayed a highly selective and sensitive response to Cu^{2+} via perceptible color and UV–vis absorbance changes among the other tested metal ions. In the presence of Cu^{2+}, the peak at 341 nm disappeared and concomitantly a new charge transfer absorption band emerged at 381 and 450 nm. Moreover, the compound formed host-guest complexe in 1:1 stoichiometry.

1. Introduction

Development of chemosensors for the selective recognizing and sensing of bioactive cations is one of the most burgeoning areas in current research because of their different analytical applications in biological, industrial and environmental processes [1,2]. Cu^{2+} is known as a catalytic cofactor in a variety of metallo-enzymes which includes superoxide dismutase, cytochrome coxidase, and tyrosinase etc. [3]. However, despite its biological significance, imbalance in its required concentration can lead to serious detrimental effects in the human body. The over accumularion of Cu^{2+} in humans is responsible for a lot of neurological diseases such as Alzheimer's, Wilson's and prion disease [4,5]. Owing to the importance and hazardous roles played by Cu^{2+} ions, numerous efforts have been undertaken to develop efficient and selective methods to assess Cu^{2+} ions in cells and organisms [6].

Using optical techniques to detect Cu^{2+} ions has resulted in fruitful work in the past several decades [7]. However, most of these techniques depend on fluorescent sensors due to their sensitivity [8]. Compared to fluorometric sensors, colorimetric sensors have attracted much attention for allowing so-called "naked-eye" detection in a straightforward and inexpensive manner, offering qualitative and quantitative information without using expensive equipment.

To date, a number of photochromic compounds as versatile sensors for ions have been reported [9]. As one of the most promising photoresponsive materials, diarylethenes are especially attractive due to their excellent thermal stability, remarkable fatigue resistance, and rapid response [10]. Most chemosensors based on diarylethene derivatives were obtaind by introduce an ion recognizing group into the molecular structures [11-13]. Herein, a colorimetric chemosensor based on a novel diarylethene with 8-hydroxyquinoline-linked salicylhydrazide Schift base has been designed and synthesized, in which act as a recognition unit. It showed selective colorimetric and UV–vis spectral changes in the presence of Cu^{2+} in CH_3CN. The photochromism of **1o** is shown in Scheme 1.

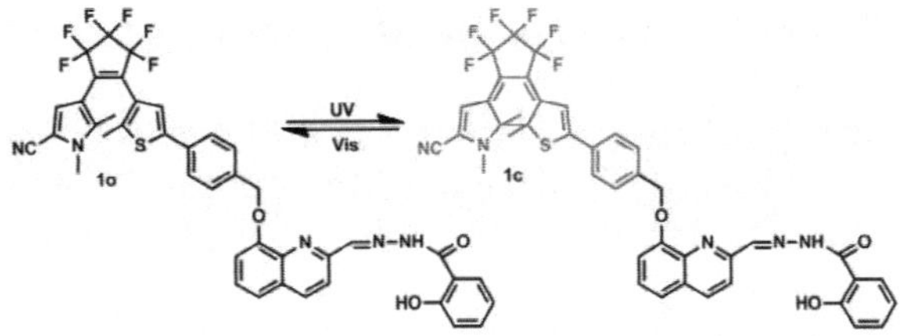

Scheme 1. Photochromism of diarylethene **1.**

2. Experimental

2.1. General

NMR spectra were collected on a Bruker AV400 (400 MHz) spectrometer by using tetramethylsilane (TMS) as an internal standard and with $CDCl_3$ as the solvent. Elemental analysis was carried out with a PE CHN 2400 analyzer. UV/Vis spectra were recorded on an Agilent 8453 UV/Vis spectrophotometer. TLC plates were visualized by 254 nm UV irradiation. Fluorescence spectra were recorded with a Hitachi F-4600 fluorescence spectrophotometer. Photo-irradiation experiments were performed using an SHG-200 UV lamp, CX-21 ultraviolet fluorescence analysis cabinet, and a BMH-250 visible lamp. Mass spectra were performed using on Bruker AmaZon SL spectrometer. The solutions of metal ions (0.1 mol L^{-1}) were obtained by the dissolution of their respective metal nitrates in distilled water, except for Mn^{2+}, Hg^{2+}, K^{+}, and Ba^{2+} (their counter ions were chloride ions). Chemical reagents were purchased from either Alfa or TCI and used without further purification. Solvents were from Beijing Chemical Works. Anhydrous solvents were of spectro-grade and purified by distillation prior to use. All solution-phase reactions were performed under an atmosphere of dry argon or nitrogen.

2.2. Synthesis

The synthetic route to the target compound was shown in Scheme 2. **3-5** were synthesized by the similarily reported methods [14,15].

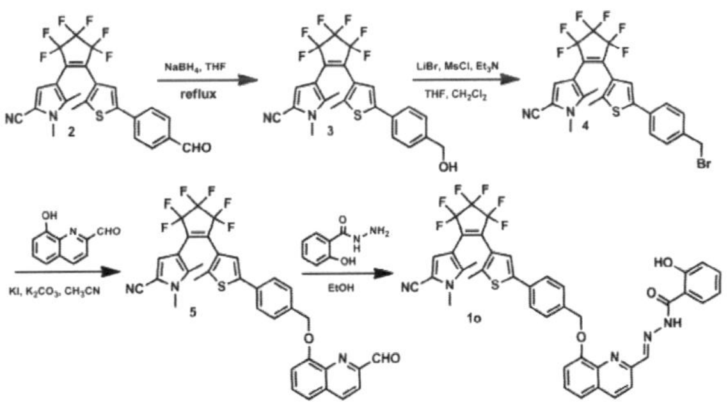

Scheme 2. Synthesis route to diarylethene **1o**.

*2.2.1. Synthesis of 1-[2-methyl-5-(4-hydroxymethylphenyl)-3-thienyl]-2-(2-cyano-1,5-dimethyl-4-pyrryl)perfluorocyclopentene (**3**).* To a stirred solution of compound **2** (2.47 g, 5.00 mmol) in THF (50 mL), $NaBH_4$ (0.27 g, 7.00 mmol) was added and continuously stirred. After refluxing for 2 hrs, the mixture was cooled to room temperature and quenched with water (40 mL). The mixture was extracted with ethyl acetate, dried over $MgSO_4$, filtered, and evaporated. The residue was purified by chromatography on silica gel using the mixture solvent of petroleum ether and ethyl acetate (v/v = 2:1) as the eluent to give **3** (2.16 g, 4.35 mmol) as a light yellow solid in 87% yield. ^{1}H NMR ($CDCl_3$, 400 MHz), δ (ppm): 1.77 (s, 3H), 1.98 (s, 3H), 3.60 (s, 3H), 4.71 (s, 2H), 6.93 (s, 1H), 7.23 (s, 1H), 7.38–7.35 (m, 2H), 7.53–7.55 (m, 2H).

*2.2.2. Synthesis of 1-[2-methyl-5-(4-bromomethylphenyl)-3-thienyl]-2-(2-cyano-1,5-dimethyl-4-pyrryl)perfluorocyclopentene (**4**).* To a stirred anhydrous CH_2Cl_2 (40 mL) of compound **3** (0.99 g, 2.00 mmol) was added MsCl (0.26 ml, 3.20 mmol), TEA (0.50 ml, 3.60 mmol) with vigorous stirring in the cold water bath. Stirring was continued for 5 min at 273 K, 10 mL anhydrous THF containing LiBr (0.69 g, 8.00 mmol) was added and the reaction mixture was stirred for 30 min. at this temperature. The reaction was warmed to room temperature and stirred for 12 hrs. The product was extracted with ethyl acetate. The organic layer was washed with saturated sodium bicarbonate and water, then dried over $MgSO_4$, filtrated, and evaporated. The crude product was purified by column chromatography on silica gel using the mixture solvent of petroleum ether and ethyl acetate (v/v = 10:1) as the eluent to obtain compound **4** (1.07 g, 1.82 mmol) as a light yellow solid in 88% yield. ^{1}H NMR ($CDCl_3$, 400 MHz), δ (ppm): 1.76 (s, 3H), 1.96 (s, 3H), 3.61 (s, 3H), 4.52 (s, 2H), 6.94 (s, 1H), 7.25 (s, 1H), 7.41–7.43 (m, 2H), 7.50–7.53 (m, 2H).

*2.2.3. Synthesis of 1-[2-methyl-5-(2-formylquinoline-8-methoxyl)]methyl]-3-thienyl]-2-(2-cyano-1,5-dimethyl-4-pyrryl)perfluorocyclopentene (**5**).* To a stirred solution of **4** (1.12 g, 2.00 mmol) in anhydrous acetonitrile (30 mL), 8-hydroxyquinoline-2-carboxaldehyde (0.28 g, 1.60 mmol), K_2CO_3 (0.66 g, 4.00 mmol), and KI (0.80 g, 2.20 mmol) were added and continuously stirred. After refluxing for 6 h, the mixture was cooled to room temperature and concentrated under vacuum. The crude product was purified by column chromatography on silica gel using petroleum ether/ethyl acetate (v/v = 6/1) as the eluent to afford **5** as a yellow solid in 70%. ^{1}H NMR ($CDCl_3$, 400 MHz), δ (ppm): 1.70 (s, 3H), 1.89 (s, 3H), 3.54 (s, 3H), 5.43 (s, 2H), 6.86 (s, 1H), 7.08–7.10 (m, 1H), 7.18 (s, 1H), 7.41–7.48 (m, 6H), 8.01–8.03 (m, 1H), 8.22–8.24 (m, 1H), 10.26 (s, -CHO).

*2.2.4. Synthesis of 1-[2-methyl-5-(2-salicylhydrazonephenylquinoline-8-methoxyl)]methyl]-3-thienyl]-2-(2-cyano-1,5-dimethyl-4-pyrryl)perfluorocyclopentene (**1o**).* **5** (0.33 g, 0.50 mmol) and salicylhydrazide (0.085 g, 0.55 mmol) were added in ethanol (10 mL). The mixture was heated at 353 K and was refluxed for 12 h until no **5** was detected by the TLC silica gel plate. The mixture was cooled to room temperature and concentrated under vacuum. The crude product was purified by column chromatography on silica gel using ethanol as the eluentrecrystallization to give **1o** 0.24 g as a yellow solid in 61% yield. Anal. calcd for $C_{41}H_{29}F6N_5O_3S$ (%): calcd C, 62.57; H, 3.72; N, 8.91. Found C, 62.53; H, 3.70; N, 8.89. ^{1}H NMR ($CDCl_3$, 400 MHz), δ (ppm): 1.78 (s, 3H), 1.95 (s, 3H), 3.60 (s, 3H), 5.51 (s, 2H), 6.60–6.64 (t, 1H, J = 8.0Hz), 6.91–6.95 (m, 2H), 7.09–7.12 (m, 1H), 7.20 (s, 1H), 7.29–7.30 (m, 1H), 7.37–7.39 (m, 2H), 7.45–7.48 (m, 4H), 7.63–7.65 (d, J = 4.0Hz, 1H), 7.83 (s, 1H), 8.17–8.19 (m, 1H), 8.33–7.35 (d, J = 4.0Hz, 1H), 12.31 (s,1H),15.81 (s, 1H).

2.3. Procedures for metal-ion sensing
Stock solutions of the metal ions (0.1 mol L^{-1}) were prepared in deionized water. A stock solution of **1o** (1.0×10^{-3} mol L^{-1}) was prepared in CH_3CN, and was then diluted to 2.0×10^{-5} mol L^{-1}. Titration experiments were performed by placing 2 mL of **1o** in a quartz cuvette of 1 cm optical path length, and then adding the Cu^{2+} stock solution incrementally by means of a micro-pipette. Spectra were recorded 3 sec after the addition. Test samples for selectivity experiments were prepared by adding appropriate amounts of metal ion stock solutions to 2 mL of **1o**. In competition experiments, Cu^{2+} was added to solution containing **1o** and the other metal ions of interest.

2.4. Determination of the stoichiometry of the 1o-Cu^{2+} complex
According to the method for continuous variation. A series of solutions of **1o** and Cu^{2+} at ratios of 1:9, 2:8, 3:7, 4:6, 5:5, 6:4, 7:3, 8:2 and 9:1 were prepared, and the absorption spectra were recorded. When plot the fluorescence intensity against the [Cu^{2+}] / ([Cu^{2+}] + [**1o**]), a feature point was obtained at 0.5 in abscissa, indicating the 1:1 stoichiometry of the **1o**-Cu^{2+} complex.

3. Results and discussion

The absorption spectral and color changes of **1o** by photoirradiation in acetonitrile are shown in Fig. 1A. Upon irradiation with 297 nm light, a new absorption band centered at 619 nm emerged due to the formation of the closed-ring isomer **1c**, accompained with a color change from colorless to blue. The blue color faded completely to colorless upon irradiation with visible light (λ > 510 nm), and the absorption spectrum returned to the initial state of **1o**. When the photostationary state was arrived, a clear isosbestic point was observed at 317 nm, indicating that the process was in agreement with a two-component photochromic reaction [16]. The fluorescence property of **1o** was measured in acetonitrile (2.0 × 10^{-5}mol L^{-1}) at room temperature (Fig. 1B). When excited with 344 nm light, the fluorescent emission peak of **1o** appeared at 472 nm. Upon irradiation with 297 nm light, the emission intensity decreased due to the formation of weakly fluorescent closed-ring isomers of **1c** [17]. The back irradiation with appropriate visible light (λ > 500 nm) regenerated the open-ring isomer of **1o**, and recovered the original emission intensity. The emission intensity of **1** was quenched to ca. 33% in the photostationary state.

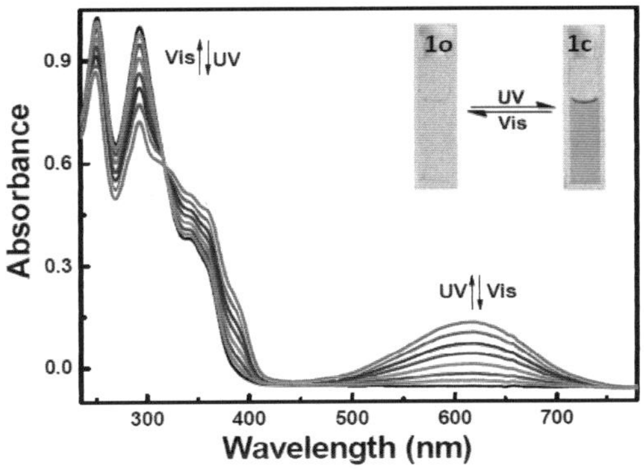

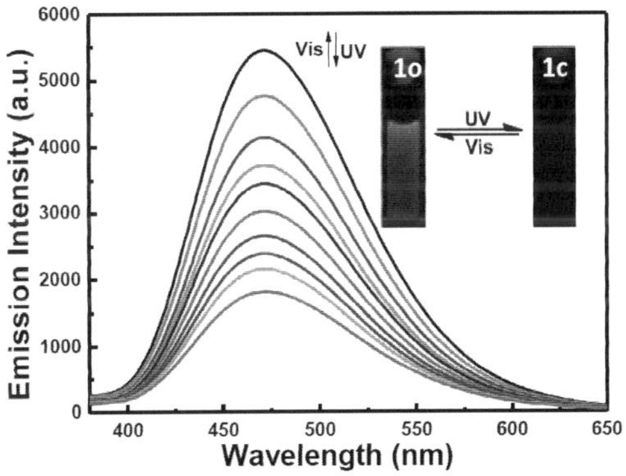

(A) (B)

Figure 1. The absorption and emission spectral changes of **1o** by photoirradiation in acetonitrile (2.0 × 10^{-5}mol L^{-1}): (A) absorptionspectral change, inset: color changes by stimui of 297 nm and visible light (λ > 500 nm); and (B) emission spectral change.

The naked-eye detectable color change of **1o** (2.0 × 10^{-5} mol L^{-1}) was observed by adding an equivalent amount of various cations such as (K^+, Ca^{2+}, Mg^{2+}, Ba^{2+}, Sr^{2+}, Zn^{2+}, Cd^{2+}, Ni^{2+}, Mn^{2+}, Co^{2+}, Pb^{2+}, Cr^{3+}, Al^{3+}, Fe^{2+} and Cu^{2+}). **1o** showed a perceptible color change from colorless to yellow in the presence of Cu^{2+}. In UV–vis absorption study (Fig. 2), **1o** alone showed two absorption bands at 252 nm, 293 nm, and 341 nm due to π–π* transitions [18]. Upon addition of Cu^{2+}, the peak at 341 nm disappeared and concomitantly a new charge transfer absorption band appeared between 358 and 550 nm with absorption maxima at 381 and 450 nm. No noticeable color and spectral changes were observed in the presence of other tested metal ions. The charge transfer promoted color and spectral changes presumably occurred due to the delocalization of electrons from the quinoline-N, imine-N and salicylhydrazide-OH to Cu^{2+} during complexation.

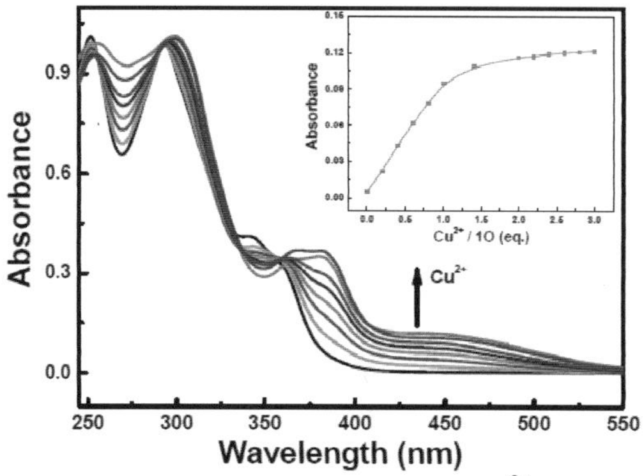

Figure 2. Absorption spectra changes of **1o** upon addition of Cu^{2+} in acetonitrile (2.0×10^{-5} mol L^{-1}). Inset: the change of absorbance at 450 nm with the addition of different equiv Cu^{2+}.

For competition studies, **1o** was treated with one equivalent of Cu^{2+} ion in the absence and presence of one equivalent of other metal ions (Zn^{2+}, Al^{3+}, Cr^{3+}, Co^{2+}, Mn^{2+}, Hg^{2+}, Pb^{2+}, Cd^{2+}, K^{+}, Ca^{2+}, Fe^{3+}, Mg^{2+}, Ba^{2+}, Sr^{2+} and Ni^{2+}) (Fig. 3). No significant interference was observed in the detection of Cu^{2+} ion in the presence of other metal ions.

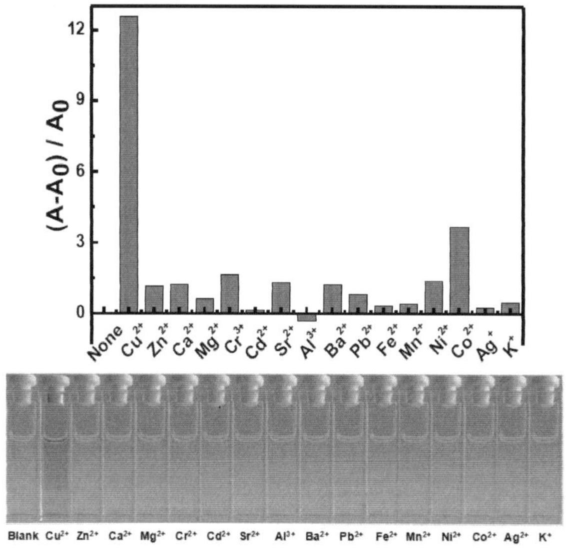

(A) (B)

Figure 3. Competitive tests for the absorbance responses of **1o** to various metal ions: (A) absorptionspectral change (Bars represent the ratio ofabsorbance at 450 nm), and (B) color change of **1o** upon addition of different cations in acetonitrile (2.0×10^{-5} mol L^{-1}).

The binding ability of **1o** towards Cu^{2+} was estimated by performing UV–vis absorption titration experiment. Upon successive addition of an incremental amount of Cu^{2+} to the solution, the absorption band was gradually red shifted with the appearance of an isosbestic point at 358 nm indicative of the formation of a single species in solution. Using Job's plot (Fig. 4), keeping the sum of the initial concentration of Cu^{2+} and **1o** at 2.0×10^{-5} mol L^{-1} and the changes of Cu^{2+} molar ratio from 0 to 1, the stoichiometry of the **1o** and Cu^{2+} was obtained. When excess amounts of Cu^{2+} were added to **1o**, a new peak at 786.1967 *m/z* that was assigned to **1o**–Cu^{2+} with a 1:1 binding mode. According to the literature [19], the possible binding mode was shown in Fig. 5.

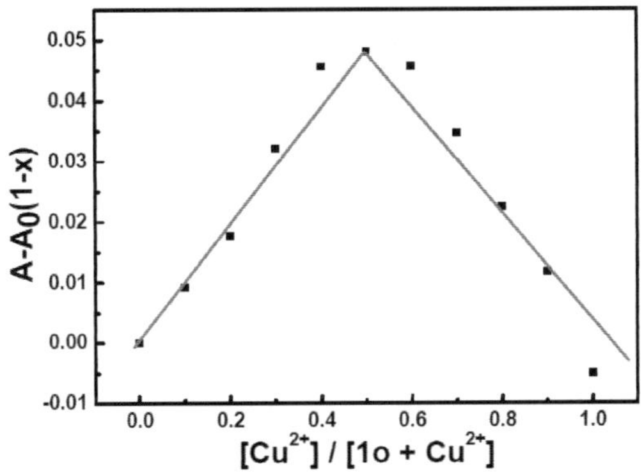

Figure 4. Job's plot showing the 1 : 1 complex between Cu^{2+} and **1o**.The total concentration of Cu^{2+} and **1o** was 2.0×10^{-5} mol L^{-1}.

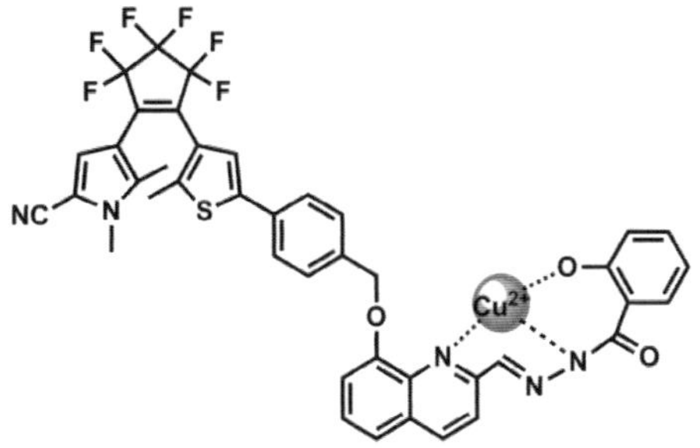

Figure 5. The possible binding mode between Cu^{2+} and **1o**.

In summary, a new easy-to prepare and chemosensor derived from a diarylethene for the selective detection of Cu^{2+} was designed and synthesized. The color of the sensor turned from colorless to yellow with the appearance of a new charge transfer band at 381 and 450 nm selectively in the presence of Cu^{2+} which allowed to detect this metal ion. The result is helpful for the design of novel sensor based on photofunctional materials.

References

[1] M. Chhatwal, A. Kumar, V. Singh, R.D. Gupta, S.K. Awasthi, Coord. Chem. Rev. **292**, 30 (2015)
[2] G.R.C. Hamilton, S.K. Sahoo, S. Kamila, N. Singh, N. Kaur, B.W. Hyland, J. F. Callan, Chem. Soc. Rev. **44**, 4167 (2015)
[3] R. Uauy, M. Olivares, M. Gonzalez, Biomed. Pharmacother. **57**, 134 (2003)
[4] U.N. Yadav, P. Pant, S.K. Sahoo, G.S. Shankarling, RSC Adv. **4**, 42647 (2014)
[5] U. Fegade, A. Saini, S.K. Sahoo, N. Singh, R. Bendre, A. Kuwar, RSC Adv. **4**, 39639 (2014)
[6] K.P. Carter, A.M. Young, A.E. Palmer, Chem. Rev. **114**, 4564 (2014)
[7] X.Q. Chen, T. Pradhan, F. Wang, J.S. Kim, J.Y. Yoon, *Chem. Rev.***112**, 1910 (2012)
[8] Y.E. Spichiger-Keller, Wiley-VCH, New York, 1998
[9] M. Natali, S. Giordani, *Chem. Soc. Rev.* **41**, 4010 (2012)
[10] M. Irie, T. Fukaminato, K. Matsuda, S. Kobatake, Chem. Rev. **114**, 12174 (2014)
[11] S.Z. Pu, H.C. Ding, C.H. Zheng, H.Y. Xu, J. Phys. Chem. C **118**, 7010 (2014)
[12] H. Tian, B. Qin, R. Yao, X. Zhao, Adv. Mater. **15**, 2104 (2003)
[13] S.J. Xia, G. Liu, S.Z. Pu. J. Mater. Chem. C **3**, 4023(2015)
[14] Liu, G.; Pu, S. Z.; Wang, X. M.; Liu, W. J.; Yang, T. S. *Dyes Pigm.* **90**, 71 (2011)
[15] S.Z Pu,, L.L. Ma, G. Liu, H.C. Ding, B. Chen, *Dyes Pigm.* **113**, 70 (2015)
[16] T. Nakashima, K. Miyamura, T. Sakai, T. Kawai, Chem. Eur. J. **15**, 1977 (2009)
[17] T. Kawai, M.S. Kim, T. Sasaki, *Opt. Mater.* **21**, 275 (2003)
[18] Z.X. Li, L.Y. Liao, W. Sun, C.H. Xu, C. Zhang, C.J. Fang, C.H. Yan, J. Phys. Chem. C **112**, 5190 (2008)
[19] Y.C. Liu, Z.Y. Yang, J. Organomet. Chem. **694**, 3094 (2009)

Acknowledgment

The authors are grateful for the financial support from the National Natural Science Foundation of China, 51373072), the Young Scientist Training Program of the Jiangxi Province (20153BCB23008), and the JXSTNU Sci-Tech innovation team (2015CXTD002).

IOP Conf. Series: Materials Science and Engineering **222** (2017) 012020 doi:10.1088/1757-899X/222/1/012020

Review of 72.5kV double-break vacuum circuit breaker based on rapid repulsion actuator

Tang Zhuofan[1,a], Qin Lijun[1]

[1]Electrical Engineering, North China Electric Power University, 102206 Beijing, China

E-mail: [a] tangvera@126.com

Abstract. 72.5kV double-break vacuum circuit breakers based on rapid repulsion actuator remain blank in China. Based on the theoretical analysis and experimental results from researchers, the design of 72.5kV double-break vacuum circuit breakers based on rapid repulsion actuator was presented. It takes the form of double-break, using two standard 40.5kV vacuum interrupter in series at the bottom, which adopt a permanent magnetic repulsion actuator. The permanent magnetic repulsion actuator consists of rapid repulsion actuator and magnetic retentivity actuator. On the basis above, we produced the prototype, and the superiority of the design was verified through the experiments.

1. Introduction

The rapid development of China's power system greatly promoted the development of high-voltage vacuum circuit breakers. Especially in recent years, because of the research and application of the multi-power and non-disturbance fast switching device for high voltage power network, the action speed of the high-speed switch has been put forward higher requirements. At present, the relevant units at home and abroad has made progress in the development of 110kV high voltage vacuum circuit breakers, such as the breaking performance and the mechanism design of the vacuum interrupter chamber. But the 72.5kV double-break vacuum circuit breakers based on rapid repulsion actuator remain blank in China. Independent research and development should be made to the related key technologies and the equipments[1].

Because there is a saturation effect between the break voltage of the breaker and the gap length of the fracture, when the length of the vacuum gap is longer than usual, simply increasing the length of the vacuum gap can't improve the pressure level of vacuum circuit breakers. When the relatively long vacuum gap is divided into several relatively short vacuum gaps, and put them in series, you can improve its pressure level. Actually, this is the basis of using the multi-break vacuum circuit breaker technology to obtain high-voltage circuit breakers. In order to promote the vacuum circuit breaker to a higher voltage level of development and application, using the advantages of multi-break vacuum circuit breaker is a more sensible choice.

Due to the stray capacitance, the partial pressure of the multi-break circuit breaker is not uniform, which will make the breaking performance of circuit breakers greatly worse. Therefore, this paper also discusses the approach of the domestic and foreign researchers using on the fracture of the pressure equalization.

Based on the theoretical analysis and experimental results from researchers, the design of 72.5kV double-break vacuum circuit breakers based on rapid repulsion actuator was presented. It takes the form of double-break, using two standard 40.5kV vacuum interrupter in series at the bottom, which adopt a permanent magnetic repulsion actuator. The permanent magnetic repulsion actuator consists of rapid repulsion actuator and magnetic retentivity actuator. On the basis above, we produced the prototype, and the superiority of the design was verified through the experiments.

2. Theoretical basis of 72.5kV double-break high voltage vacuum circuit breaker

2.1. Pressure saturation effect of vacuum gap

Vacuum gas molecules in the air gap is minimal. As the breakdown of vacuum air gap by a variety of factors, the physical process is very complex[2]. After researchers repeated theoretical analysis and experimental demonstration several times,the relationship between the breakdown voltage and the vacuum gap length is obtained[3]. When the length of the vacuum gap is relatively small, generally less than 0.5cm, the breakdown voltage will proportionally increase as the length of the vacuum gap increases. When the vacuum gap length is large, generally more than 1cm, the breakdown voltage does not increase proportionally as the length of the vacuum gap increases,showing a pressure saturation characteristics. The relationship is:

$$U = \begin{cases} Kd, & d \le 0.5cm \\ Kd^{\alpha}, & d \ge 1.0cm \end{cases} \tag{1}$$

Where U is the breakdown voltage in kV; K is the proportional constant; the exponential factor α is usually in the range of 0.4 to 0.7; d is the length of the vacuum gap in cm.

Physical principles of multi-break vacuum circuit breaker, in essence, is the use of multiple small vacuum gap in series. Then make full use of the advantages of the short vacuum gaps. So we will obtain a vacuum circuit breaker of higher tolerance voltage level. Under ideal circumstances, the researchers assumed that the voltage of each interrupter in a multi-break vacuum circuit breaker was evenly distributed. The relationship between the breakdown voltage and vacuum gap length is :

$$U_n = nKd^{\alpha} \tag{2}$$

Where Un is the multi-break vacuum circuit breaker breakdown voltage. The meaning of the other physical quantities is the same as above. Under ideal circumstances, assuming that the interrupter vacuum gap length of a single-break vacuum circuit breaker is n times the interrupter vacuum gap length of a multi-break vacuum circuit breaker. The breakdown voltage of single-break vacuum circuit breaker is:

$$U_1 = K(nd)^{\alpha} \tag{3}$$

The gain multiplies of breakdown voltage of the multi-break vacuum circuit breaker are:

$$K_n = \frac{U_n}{U_1} = n^{(1-\alpha)} \tag{4}$$

As can be seen from equation (4), for a double-break rapid vacuum circuit breaker, if the exponential factor is 0.5, the breakdown voltage gain factor is 1.414. Under ideal circumstances, compared with single-break vacuum circuit breakers, multi-break vacuum circuit breakers have higher pressure level. It can be seen that multi-break rapid vacuum circuit breakers have the following advantages: The voltage applied to each of the break ports can be reduced by the configuration of a plurality of ports, and the arc-gap recovery voltage of each port can be reduced. The arc is divided into a plurality of small arc segments by a plurality of fractures, then connect them in series. As the same contact stroke, the tensile arc of multi-break structure is longer than the single break structure, which is mean that arc gap resistance becomes larger. Circuit breaker opening speed has been improved, while accelerating the recovery speed of the media. Circuit breaker control mechanism has reduced mechanical energy, laying the foundation for its manufacturing.

IOP Conf. Series: Materials Science and Engineering **222** (2017) 012020 doi:10.1088/1757-899X/222/1/012020

2.2. Dynamic properties of the multi-break vacuum circuit breaker

The insulation properties and changes of the multi-break vacuum circuit breakers in a variety of practical situation are called the dynamic insulation properties of the multi-break rapid vacuum circuit breaker, including the medium strength and the post-arc re-breakdown ability of the circuit breaker in a high-current breaking. That is reflecting the ability of the vacuum breaker to break current. The dynamic insulation characteristics of the multi-break fast vacuum circuit breaker are closely related to the reliability of the circuit breaker operating safely and stable in power system. So it has been the focus of researchers.

In paper [3], the breaking performance of the double-break circuit breaker was tested by using two single-break vacuum circuit breakers. Using the arc time to reflect the circuit breaker breaking capacity, it is concluded that the breaking performance of the double-break vacuum circuit breakers are better than the single-break fast vacuum circuit breakers. In paper [4], two 24kV fast vacuum circuit breakers are connected in series, the experimental results are analyzed. It is found that the breaking performance of the double-break vacuum circuit breakers is 1.3 times higher than that of the single-break vacuum circuit breakers.

3. Fracture pressure equalization technique of 72.5kV double-break vacuum circuit breakers

Due to the stray capacitance, the fracture partial pressure of the multi-break circuit breaker is not uniform. The way by increasing the installed equalizing capacitor is mainly taken at home and abroad, thereby enhancing the uniform distribution properties of the multiple-break fast breaker's fracture voltage. In paper [5], the breaking time of the circuit breaker is used to characterize its breaking performance. It is proved by experiments that increasing the voltage equalizing capacitor has a positive effect on improving the breaking performance of the double-berak vacuum circuit breakers. Throughout the course of the experiment, when no additional installation of voltage equalizing capacitor, the successful breaking of the double-berak vacuum circuit breaker was happened 2.8ms after the arcing. At the same time, there is no clear dispersion region in the whole experiment. After equipping with a 100pF equalizer capacitor, there is a clear dispersion region in the whole experiment. In particular, when the arc time beyond 2.3ms, each fracture of the double-berak vacuum circuit breaker can successfully break.

In paper [6], the experimental results show that the distribution of the recovery voltage on each fracture is more uniform when there is a equalizer capacitor. But the capacity of the equalizer capacitor is not the bigger the better. Because it may cause that the arc time of two vacuum interrupter is different, leading to the breakdown of the double-berak vacuum circuit breaker. So the breaking performance of the double-berak vacuum circuit breakers is deteriorated.

In paper [7], the positive impact of the installation of the equalizer capacitors on the multi-break fast break circuit breakers and the negative issues followed by the installation is comprehensively overviewed. At the same time, the theory of the positive influence and the new problem is studied.Some existing theoretical and experimental research limitations are put forward. And the preliminary analysis results are given. Based on the above theoretical research, the author gives some constructive suggestions on the reasonable value of equalizing capacitance of multi-break vacuum circuit breakers, playing a positive role in the further development of the equalizer capacitors of the multi-break vacuum circuit breakers. In paper [8], the negative effects caused by using the equalizer capacitors are demonstrated in theory and experiment.

In paper [4], in order to avoid the risk of the installation of equalizing capacitor, a new design of the double-break circuit breakers is obtained. The rational use of fracture structure is arranged so as to optimize the allocation of two fracture partial pressure. Because the connection distance between the two vacuum interrupters is very small and the position at which the fracture is connected is far away from the ground, generally more than 1000mm, the port dispersion capacitance of the conductor at the junction is very small, which can be ignored. So its equivalent circuit can be shown in Figure 1.

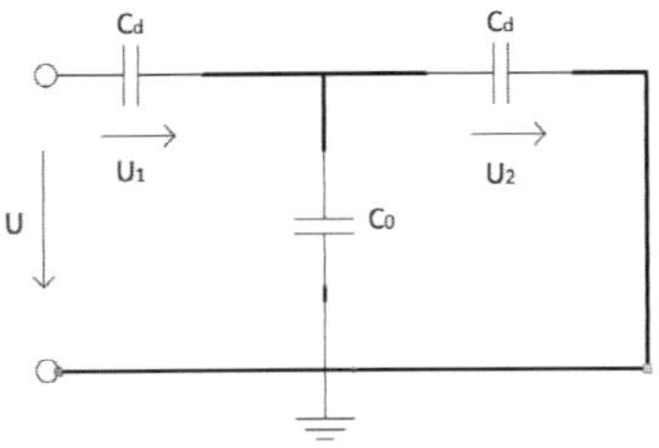

Figure 1. The equivalent circuit of the new double-break vacuum breaker.

Where U is the supply voltage; U_1 and U_2 are the fracture voltage; C_d is the fracture capacitance, usually 1000pF to 2000pF; C_0 is the capacitance to the ground,usually 20pF to 60pF. It can be seen that, in this form of structure, the fracture voltage is basically the same, because C_d is far greater than C_0.

4. Structure design of the double-break high voltage vacuum circuit breaker

4.1. Appearance structure design of 72.5kV double-break vacuum circuit breaker

When designing the appearance structure of the double-break vacuum circuit breaker, the "H" type structure and inverted "A" type structure are generally used. The drive mechanism of the inverted "A" type fast vacuum circuit breaker must be equipped with a rotating part to turn the force of vertical movement to a certain angle oblique force. Not only will it affect the circuit breaker opening and closing speed, but also affect the operating force. As can be seen from the above, using "H" type structure can make full use of its structural advantages to achieve the fracture pressure. In this paper, the form of double-break is taken, and two standard 40.5kV vacuum interrupter are used in series at the bottom. It can be shown below in Figure 2.

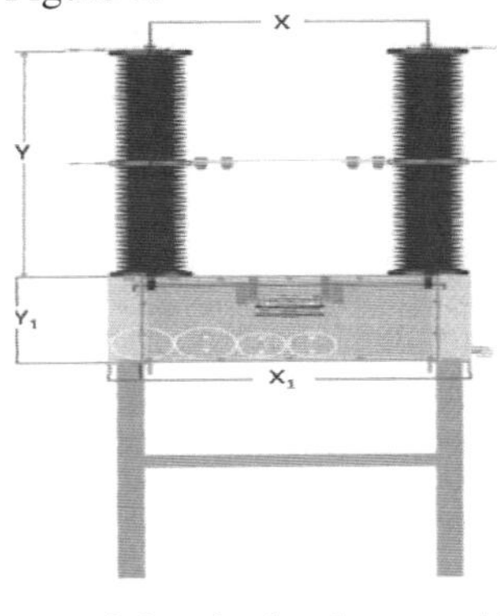

Figure 2. Appearance of the single-phase rapid vacuum breaker.

In Figure 2, X and Y are designed in accordance with the 72.5kV voltage level. X_1 and Y_1 are the drive dimensions that match the 72.5kV vacuum circuit breaker. And sealing should be considered for outdoor installation. Considering the insulation standard of relevant voltage grade, the final design dimension of X is 650mm, Y is 1050mm, X_1 is 850mm, Y_1 is 400mm.

4.2. Actuating mechanism structure design of 72.5kV double-break vacuum circuit breakers

Overview *of actuator*. The main performance of the rapid vacuum circuit breaker is directly presented in the opening and closing action time and speed of the static and dynamic contact of the circuit breaker. Opening and Closing action must be completed directly by the operating mechanism. Therefore, the performance of the selected operating mechanism will directly affect the stability and reliability of the high-voltage vacuum circuit breaker[9,10].

The energy is stored in the traditional spring operating mechanism by the built-in elastic properties of the spring. And it does not require high-powered DC power as operating power. However, the mechanical structure of the device is complicated, and the production cost is high. Also the precision is low, the reliability can not be guaranteed. The electromagnetic mechanism structure drive is simple and reliable. But in order to provide greater operating power, it usually requires a larger traction electromagnet, power supply unit and large area cable. It need to be equipped with expensive battery, leading to the heavy structure and long action time. The traditional permanent magnet actuator has many problems, such as the large cumulative motion tolerance of each control link and the slowly action response. So the action of each breaker of the circuit breaker is not synchronized. Thus, the traditional multi-break vacuum circuit breaker is not widely used in practice. Therefore, the development of higher voltage single-break fast vacuum circuit breakers has become a problem that many researchers need to solve[11-14].

At present, the study of the circuit breaker operating mechanism using eddy current electromagnetic repulsion principle has been flourishing. In this way, the requirements can be completed in a very short period of time to quickly drive the load. Therefore, many researchers pay close attention to this aspect of the study. Currently many countries in the world have the corresponding development in the field of quick vacuum circuit breaker based on electromagnetic repulsion mechanism, especially in Japan, Germany and so on. They are using electromagnetic eddy current repulsion principle to design the quick circuit breaker at a series of voltage grade, and developed relative products by use of this principle. Repulsion mechanism has obvious advantages. Repulsion mechanism does not require mechanical off and locking device, so there is few sources of failure. Thus, it can safely and reliably operate for a long time[15-20]. It is particularly worth mentioning that the movement of its operating mechanism can be accurate to the microsecond level. If this technology is applied to multi-break vacuum circuit breaker, the control accuracy and synchronization of the sub-closing points of the multi-break fast vacuum circuit breaker control mechanism will increase. In China, such research is still in its infancy. But it is reported that the vacuum contactor interrupter and high-speed eddy current repulsion agency have been used in the production of rapid circuit breaker. However, the research of high speed vortex repulsion mechanism in high voltage circuit breaker has not yet appeared. So this article has a good application prospects.

The repulsion *mechanism composition of 72.5kV double-break vacuum circuit breakers.* In this paper, considering the characteristics of high voltage circuit breaker, a new design scheme is proposed for the repulsive force operating mechanism. The repulsive mechanism proposed in this paper is composed of a permanent magnet holding mechanism, a bidirectional repulsion mechanism and a vacuum interrupter. The schematic diagram of the circuit breaker is shown in Figure 3.

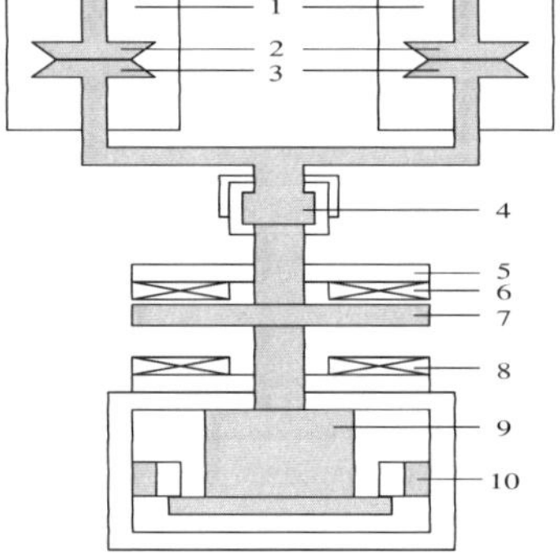

Figure 3. Appearance of the repulsion actuator of the double-break vacuum breaker.

In figure 3:

1.Vacuum interrupter: vacuum excellent insulation through the tube so that the circuit can quickly cut off the power supply and inhibit the current arc.

2.Static contact: a fixed contact that connects or disconnects a power line.

3.Moving contact: movable contact that connects or disconnects a power lines

4.Connecting rod: the whole connecting rod, used to move the moving contact action.

5.Fixed plate: used to fix opening and closing coil.

6.Opening coil: opening excitation coil, issued by the opening command to generate electricity and metal repulsion force down the repulsion.

7.Metal repulsion disk: metal copper disk, with the excitation coil to produce upward or downward repulsion, play a role in moving the moving contact.

8.Closing coil: closing excitation coil, issued a closing instruction when the power supply and metal repulsion plate upward repulsion.

9.Moving iron core: the active iron core, function with the magnet as the maintenance role.

10.Magnet: used to fix moving core.

The working principle is that the electromagnetic repulsion works as the driving force by the permanent magnet magnetic circuit in the permanent magnet to provide retention force. Self-retention is accomplished by the use of a magnet and a moving iron core structure. After opening and closing, the magnet retains the moving core by means of a magnetic force so as to achieve the self-retaining effect. The basic operation process is that the thyristor controls the charged capacitor that discharges to the coil of the repulsion mechanism, so that the coil produces a transient magnetic field. Then copper disk induces eddy current to form a reverse magnetic field, resulting in repulsive force which drives the rod movement, and then the interrupter moving contact moves, so as to achieve the effect of circuit opening and closing.

The main factors affecting the repulsive force are the thickness and the radius of the metal repulsion disk, the number of the excitation coil turn, and the parameters of initial air gap. In this paper, the parameters of the prototype components are calculated, especially the method and structure of the wound coil coil are studied deeply. Finally, the repulsion mechanism is designed. The relevant design parameters are shown in Table 1.

Table 1. The table of design parameters.

Item	Unit	Parameter
the thickness of the metal repulsion disk	mm	10
the radius of the metal repulsion disk	mm	100
initial air gap	mm	0.2
the number of excitation coil turn	turn	100
drive current	A	1500

4.3. Performance test of 72.5kV double-break vacuum circuit breaker

Test of closing and opening time. In this paper, the testing platform of the closing and opening time of the prototype is established. One channel of the dual channel oscilloscope records the pulse start time, and the other channel records the stop time. The experimental results are shown in Fig.4 and Fig.5. From the experimental results, the circuit breaker opening time of 18ms, closing time of 27ms. Therefore, the closing and opening speed of the circuit breaker prototype designed in this paper is fast. And the breaking capacity of the circuit breaker is good.

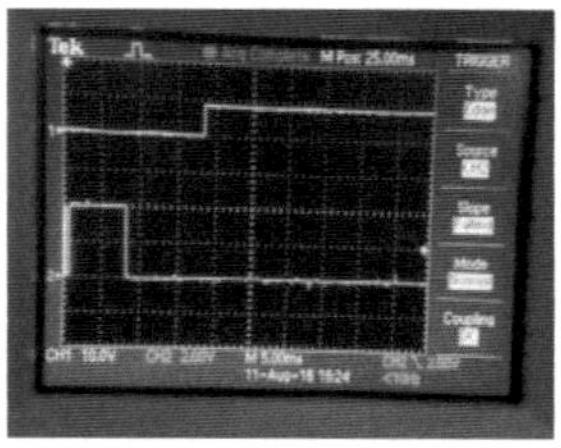

Figure 4. Closing time test.

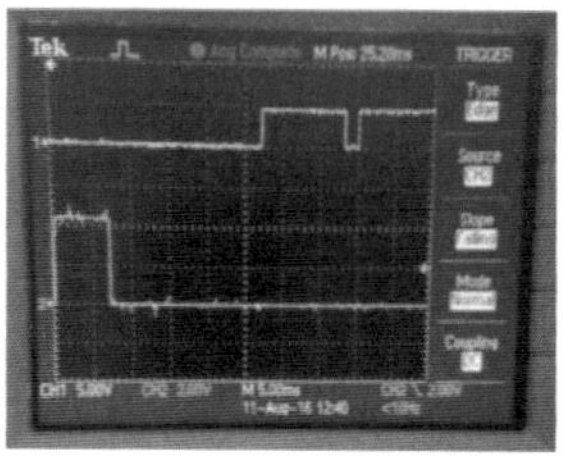

Figure 5. Opening time test.

Mechanical properties test. Connect the movable part of the linear displacement sensor to the moving lever of the circuit breaker and connect the potentiometer output voltage to the oscilloscope. By analyzing the change of the output voltage, the relationship between the stroke variation and the time of the circuit breaker is obtained. The length of the sliding rheostat is 50mm, the travel range of the breaker moving rod contact is 20mm. In this test, the charging voltage of a single charging module is 860V, when the charging modules are connected in series, the closing and opening capacitor voltage is 1700V. Then get the sub-closing circuit breaker mechanical characteristics, as shown below in Fig.6 and Fig.7.

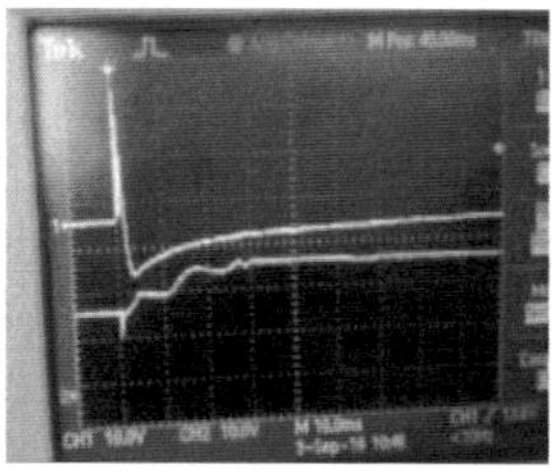

Figure 6. Closing mechanical characteristic curve.

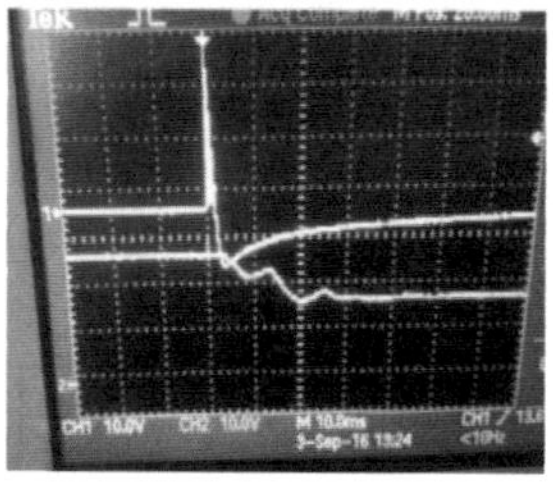

Figure 7. Opening mechanical characteristic curve.

From the figure we can see that the total time of the breaker closed from the beginning to the stability is 30ms, in fact, there is a bottom near 27ms. When open the breaker, there is a bottom near 18ms, then jump to 30ms for stability.

In summary, compared to other solutions, the circuit breaker designed in this paper has certain advantages, such as short closing and opening action time, good mechanical properties, successfully broken in many voltage level.

5. Conclusion

Based on the theoretical analysis and experimental results from researchers, the design of 72.5kV double-break vacuum circuit breakers based on rapid repulsion actuator was presented. It takes the form of double-break, using two standard 40.5kV vacuum interrupter in series at the bottom, which adopt a permanent magnetic repulsion actuator. The opening time of the circuit breaker prototype is 18ms, and the closing time of the circuit breaker prototype is 27ms.Circuit breaker can be successfully broken with a certain superiority, and the mechanical characteristics are excellent. But the structure of circuit breakers only can be applicated in the transmission grid after the type test.

The 72.5kV double-break high-speed vacuum circuit breaker developed by high-speed repulsive mechanism, can meet the high-speed requirement of high-voltage power supply without disturbance. So it has a very wide application prospect.

References

[1] Liao Mifu, Duan Xiongying, Zou Jiyan, et al. Dynamic medium recovery and statistical analysis of multi - break vacuum circuit breaker [J]. *Proceedings of the Chinese Society for Electrical Engineering*, 2003,**23** (2): 83-87.

[2] Wang Jimei. *Vacuum circuit breaker theory and its application* [M]. Xi'an: Xi'an Jiaotong

University Press, 1986.

[3] Wen Hubin, Zou Jiyan, Liao Mifu, et al.Theoretical Analysis of 126kV Double Breaking Vacuum Circuit Breaker [J]. *High-voltage electrical appliances*, 2009,**45** (2): 14-17.

[4] Fugel T， Koenig D . Pecularities of the switching performance of two 24kV vacuum interrupters in series [C]//*IEEE 19th International symposium on Discharge and Electrical Insulation in Vacuum*. Xi'an：IEEE，2000：411-414

[5] BETZ T, KOENING D. Influence of grading capacitors on the breaking capacity of two vacuum interrupters in series [C]// *Proceedings of 18th Int. Symp. on Discharges and Electrical Insulation in Vacuum*, Eindhoven, Holland:[s. n.],1998:231-238

[6] Fugel T,Koenig D.Influence of grading capacitors on the breaking performance of a 24 kV vacuum breaker series design[J]. *IEEE Trans on Dielectrics and Electrical Insulation*, 2003, **10**(4): 569-575

[7] Wu Gaobo, Ruan Jianjun, Huang Daochun, et al.Study on the study of equalizing capacitance of multi - break vacuum circuit breaker [J]. *High Voltage Apparatus*, 2011,**47** (3): 77-81.

[8] Shu Shengwen, Ruan Jiangjun, Huang Daochun, et al. Transient recovery voltage distribution mechanism and equalizing capacitance of double-break vacuum circuit breaker [J].

[9] Wen Huabin, Song Yongrui, Zou Jiyan, et al.Experimental study on design and breaking capacity of new type 126kV high voltage vacuum circuit breaker [J]. *Proceedings of the Chinese Society for Electrical Engineering*, 2011, **31**(34): 198-204.

[10] Liu Donghui, Wang Jimei, Wang Zhongyi, et al.Necessity and preliminary research on development of 126kV vacuum circuit breaker in China [J]. *High-voltage electrical appliances*, 2003,**39** (2): 26-29

[11] Ma Shaohua, Xu Jianyuan, Wang Jimei.Performance Analysis of Permanent Magnet Actuator Structure and the Possibility of Using High Voltage Vacuum Circuit Breaker [J] .*High Voltage Apparatus*, 2000,**36** (4): 19-23.

[12] Fu Wanan, Song Baoyun.Study on Permanent Magnet Mechanism of High Voltage Circuit Breaker [J]. *Proceeding of the CSEE*, 2000,**20** (8): 21-26.

[13] Zou Jiyan, Wang Ying, Dong Enyuan.Electronic Actuation Concept and Practice [J] .*High Voltage Apparatus*, 2000,**36** (5): 29-31.

[14] Lin Xin, *permanent magnet mechanism and vacuum circuit breaker* [M]. Beijing: Mechanical Industry Press, 2002.

[15] Li Bo,High-speed repulsion mechanism and fault current fast monitoring device [D]. *Dalian University of Technology* 2006.

[16] Huang Yulong, Zhang Zu'an, Wen Weijie, Gao Shutong, Cheng Tiehan, Liu Jiayu.Study on Electromagnetic Repulsive Fast Actuator in High Voltage DC Circuit Breaker [J]

[17] Liang Zhuangxian, Zhuang Jinwu, Wang Chen, Liu Luhui, Dai Chao.The Simulation Analysis and Design of High Speed Contact Mechanism Based on Electromagnetic Repulsion Principle [J] .*Techniques in Electrical Engineering*. 2011 (08).

[18] Jiang Nan, electromagnetic repulsion mechanism design methods [D]. *China Ship Research Institute* 2013.

[19] Tian Peng,A New Type of High Speed Retarding Mechanism for Permanent Magnet Retention [D]. *Dalian University of Technology* 2011.

[20] Zhao Xia, fast electromagnetic actuator topology and discharge loop optimization study [D].

IOP Conf. Series: Materials Science and Engineering **222** (2017) 012021 doi:10.1088/1757-899X/222/1/012021

The research of high voltage switchgear detecting unit

Tong Ji[1,a], Wei Xie[2], Xiaoqing Wang[1] and Jinbo Zhang[3]

[1] Changzhou Pasific Automation Technology Co.,Ltd., Changzhou City, Jiangsu Province, 213022, China

[2] Jiangsu Power Design Institute Co,. Ltd., Nanjing City, Jiangsu Province, 211102, China

3College of Internet of Things (IOT) Engineering, Hohai University, Changzhou City, Jiangsu Province, 213022, China

E-mail: [a]jitong1353@foxmail.com

Abstact. In order to understand the status of the high voltage switch in the whole life circle, you must monitor the mechanical and electrical parameters that affect device health. So this paper gives a new high voltage switchgear detecting unit based on ARM technology. It can measure closing-opening mechanical wave, storage motor current wave and contactor temperature to judge the device's health status. When something goes wrong, it can be on alert and give some advice. The practice showed that it can meet the requirements of circuit breaker mechanical properties temperature online detection.

1. Introduction

High voltage switchgear is the key role between the power generation and distribution. Its safe operation guarantees the reliability of the power grid. High voltage switchgear can play a great role in the power grid system. It can make grid work well[1].And then from the requirement for our country smart power grids construction, intelligent high voltage switchgear can monitor, think, analyze, control things and other function, all in one device. It achieves a higher level of technology.

The international power grid conference investigated and showed that control and assit part failure come after mechanical failure. 29% high voltage circuit breaker major failure and 20% minor one may be due to control and assit part[2]. The switchgear second loop adopts the traditional electromagnetic mechanical structure that can't meet the demand of modern power grids[3]. So with further automatic grids development, it must go towards intelligence.

2. Composition

Intelligent high voltage switchgear is composed of switchgear and intelligent unit, both form a combined system through sensor and controller as shown in the figure 1.Unit sends data to upper computer through IEC61850[4]. Upper computer uses expert system to analyze the device status data and estimate the device lifetime.

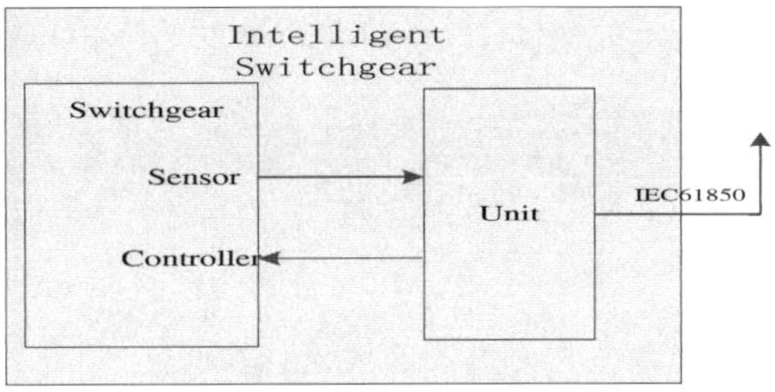

Figure 1. Intelligent Switchgear Structure Chart

The switchgear structure change that comes from the integrated design of intelligent switchgear shouldn't affect switchgear performance. The technological parameter should meet the national standards. Various sensors will be placed in and out of the switchgear connected by analog cable. The intelligent units placed in the intelligent unit cabinet should integrate control, measurement, monitor and other basic functions.

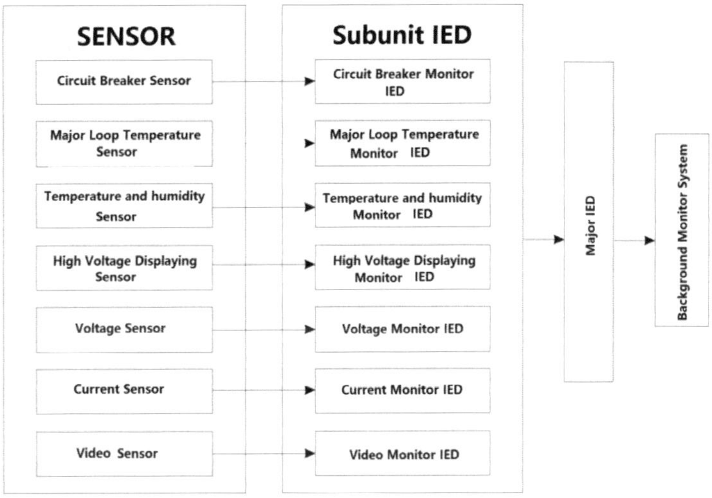

Figure 2. High Voltage Switchgear Intelligent Unit

The research of high voltage switchgear intelligent unit key technology is to study the integrated design based on the use of various sensors as shown in the figure 2.

The high voltage switchgear intelligent unit is composed of circuit breaker sensor, major loop temperature sensors, temperature and humidity sensors, high voltage displaying sensor, voltage sensor, current sensor and video sensor.

The circuit breaker sensors is used for monitoring breaker's mechanical properties, closing and opening coil current, operating stroke and speed.

The major loop temperature sensors is used for monitoring breaker's contact and busbar connector temperature.

The video sensors is used for monitoring breaker's room, busbar room and cable room.

The background monitor system has various functions such as data storage, analysis, report forms, expert decision that can pre-estimate the device life.

3. Key Point Research

3.1. Breaker's Contact and busbar Connector Temperature Sensors

The breaker's contact and busbar connector temperature sensors is composed of temperature sensor and receiver using wireless communication. The temperature sensor is installed on the breaker's contact and busbar connector. The receiver is installed in the switchgear instrument room. The structure is as shown in the figure 3.

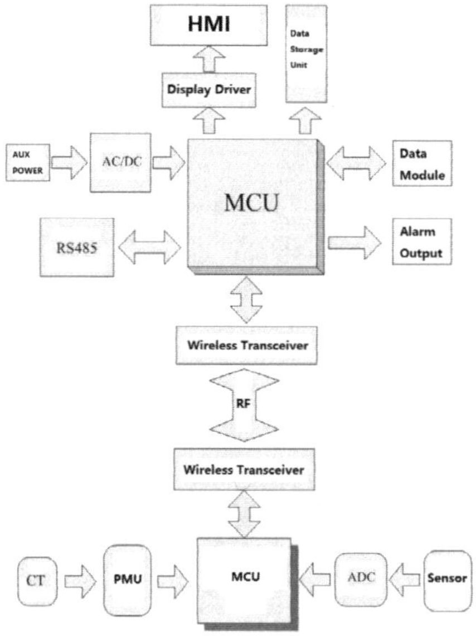

Figure 3. Major Loop Temperature Measure Principle Chart

The temperature sensor is composed of NTC resistance, CT power supply, signal conversion and transmit module as shown in the figure 4.NTC resistance put on the surface of breaker's contact closing to plum contact.NTC resistance value comes after the temperature change as shown in the figure 5.

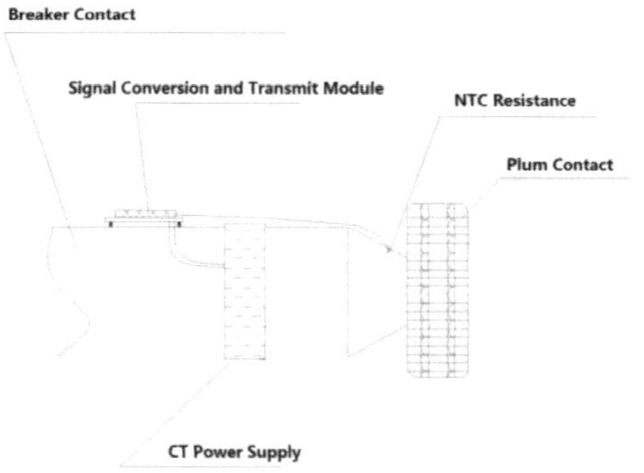

Figure 4. Temperature Sensor Principle Chart

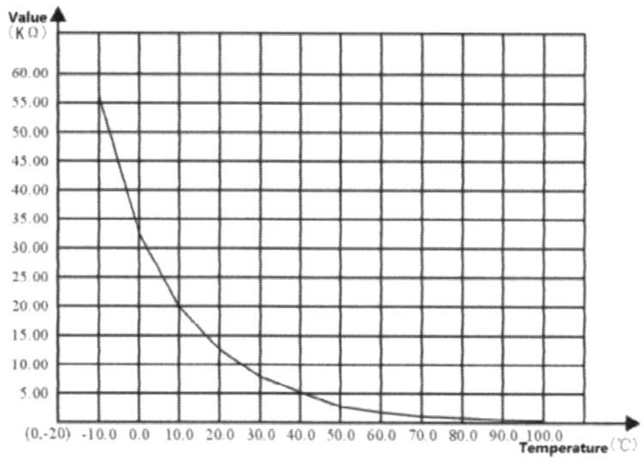

Figure 5. NTC Resistance Value-Temperature Diagram

The value of NTC resistance changes with the temperature. And it's transmitted to measuring MCU by the signal conversion and transmit module.

The signal conversion and transmit module is composed of high frequency receiving amplifier, rectifier filter, decoding circuit and alarm circuit. The wireless signal with temperature message and address codes is amplified by high frequency receiving amplifier. It's transmitted to decoding circuit and then give an alarm. The picture of real products is shown in figure 6.

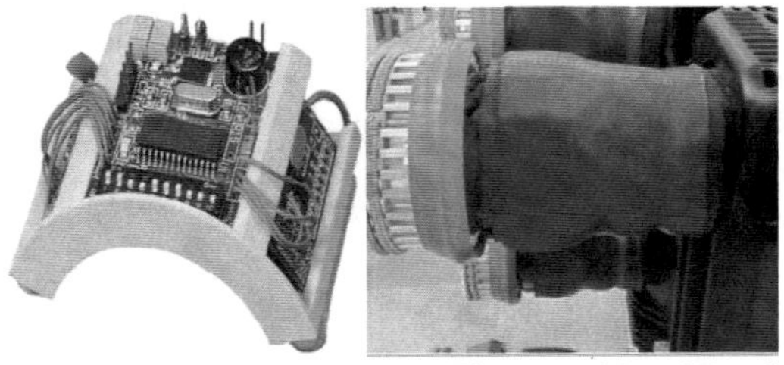

Figure 6. Temperature Sensor Picture of Real Products

3.2. Circuit Breaker's Mechanical Properties Sensor

Electromagnet is usually first level control element in the high voltage circuit breaker. The DC power supply is added to most of breakers. So there are many important mechanical error messages in the DC electromagnetic coil current waveform. The inductance value L depends on the size of coil and iron core. It has a close relation with iron core stroke S, and increases as S gets bigger. Hall sensor is used to measure the coil current, as shown in the figure 1,electromagnetic coil drive current value is 1-2A.The relation wave between coil current and time is shown in figure 7, A-electromagnet connecting rod activate, spring deformation, C\D-rod nearly arrives at deepest position, E-rod arrives at the deepest position, F-main contactor opens, G-assist contactor opens, H-rod and spring return to original position.

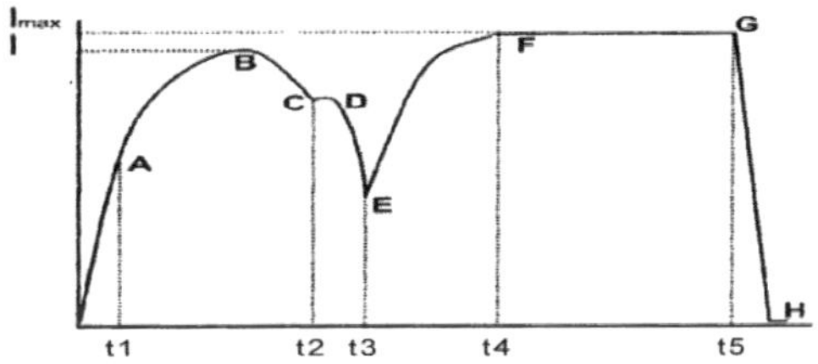

Figure 7. oil current and time wave diagram

It's shown from figure 7 that you can analyze the breaker's mechanical state from the wave. If point B becomes high, mechanism has jam error caused by the increase of resistance. If the wave disappears, the coil is broken. If point G goes left side, the action costs much more time and reduces speed[5].

The operating stroke can be detected by grating sensor, resistor sensor and so on. It chooses linear stroke sensor that is installed at linear action mechanism. It chooses rotate stroke sensor that is installed at rotate action mechanism. The output of the sensor goes by preprocess, such as isolation, clear, logical process, and you can get the breaker's operating stroke and time characteristic wave. The following parameters is calculated from the wave, average speed, average speed before closing and opening period(10ms)[6].The difficulty is that stroke sensor can not be installed in the moving contact. So it can not directly measure the stroke that is converted. Figure 8,breaker's mechanical characteristic detection block, shows that displacement sensor is installed at the bottom fixed by stand. When vacuum envelope moves, sensor moves and displacement signal changes to electrical signal.

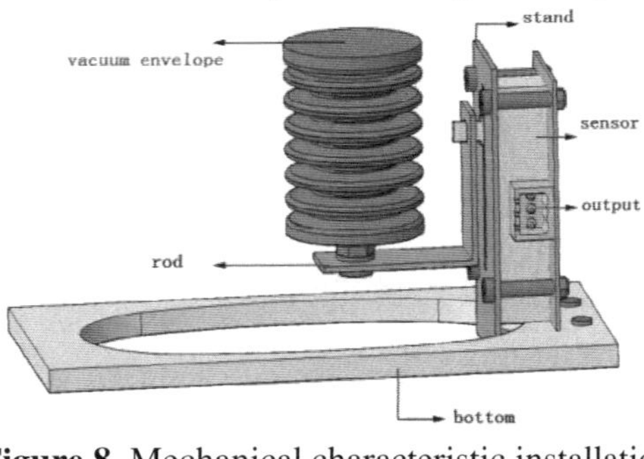

Figure 8. Mechanical characteristic installation

3.3. Intelligent Unit Embedded Software

The intelligent unit embedded software designed by the newest ARM technology made the monitoring terminal smart including CRC, the seamless coupling of terminal and secondary equipment, video monitoring and the other functions.

4. Test Result

In order to determine the feasibility of the high voltage switchgear detecting unit, a test bed is required.

Six temperature sensors are located on the surface circuit breaker's contacts,3 above and 3 below. The circuit breaker current is 1375A,power-on hours count is 6 hours. Table 1,you can compare the test unit temperature against the standard value.

Table 1.

Contact Phase	Position	Time	1h	2h	3h	4h	5h	6h
A	above	Test Unit	24	56	69	76	80	82
		Increment Value		32	45	52	56	58
		Standard	28	60	75	81	84	87
		Increment Value		32	47	53	56	59
		Difference Value	4	4	6	5	4	5
B		Test Unit	23	54	68	76	80	83
		Increment Value		21	45	53	57	60
		Standard	28	60	75	82	85	87
		Increment Value		22	47	54	57	59
		Difference Value	5	6	7	6	5	4
C		Test Unit	24	56	69	76	80	82
		Increment Value		22	45	52	56	58
		Standard	27	60	75	81	84	86
		Increment Value		23	48	54	57	59
		Difference Value	3	4	6	5	4	4
A	below	Test Unit	24	55	66	71	74	75
		Increment Value		31	42	47	50	51
		Standard	28	61	73	77	80	81
		Increment Value		33	45	49	52	53
		Difference Value	4	6	7	6	6	6
B		Test Unit	24	55	67	72	75	77

		Increment Value		31	43	48	51	53
		Standard	28	61	74	78	81	83
		Increment Value		33	46	50	53	55
		Difference Value	4	6	7	6	6	6
C		Test Unit	24	55	66	71	74	75
		Increment Value		31	42	47	50	51
		Standard	27. 7	60	72	77	79	81
		Increment Value		32	44	49	51	53
		Difference Value	4	5	6	6	5	6

From the above data analysis, test unit value is basically the same as standard value. The difference value is small and caused by the different location that the sensors are placed.

Based on the above principle high voltage circuit breaker mechanical characteristic online detection can measure opening coil current wave, closing coil current wave, closing coil broken, opening coil broken, storage coil current wave, storage time, opening time-stroke wave closing time-stroke wave. It's shown in figure 9.

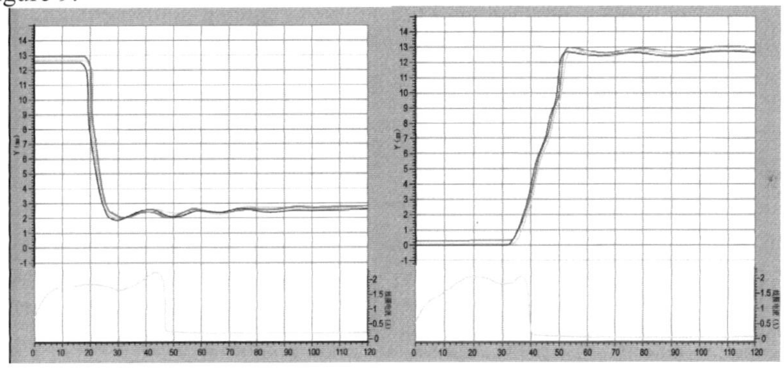

Figure 9. Mechanical characteristic wave block

Now, off line mechanical characteristic detective device uses breaker's moving contact and fixed contact voltage signal to measure breaker's iron closing and iron opening point. Based on the current signal, it can learn the iron closing and opening point, and measure the opening distance and super distance.

For the online detective device, it can use the above method, because there are high voltage when breaker's working. It can not add current signal to judge the closing and opening point. So the above parameter must be measured by the iron closing and opening point.

In order to solve the above problems, it uses the pressure sensor installed at the bottom of breaker's spring to get the state of the closing and opening iron point. But this method just suits to the new building breaker.

The mechanical closing opening stroke and electrical wave is recorded to compare with original data. If it finds ±5% deviation, it shows that breaker's mechanism has some errors and it will affect breaker's closing and opening action. It recommends that the circuit breaker should be overhauled.

5. Conclusion

Based on the above principle high voltage switchgear detecting unit can measure opening coil current wave, closing coil current wave, closing coil broken, opening coil broken, storage coil current wave, storage time, opening time-stroke wave closing time-stroke wave. For the online detective device, it can use the above method, because there are high voltage when breaker's working. It can not add current signal to judge the closing and opening point. So the above parameter must be measured by the iron closing and opening point. In order to solve the above problems, it uses the pressure sensor installed at the bottom of breaker's spring to get the state of the closing and opening iron point. It depends on the change of the pressure to judge the points. Abnormal temperature change is often the direct characterization and early warning of electrical equipment failure or accident. Through the real-time temperature monitor, it help us find high voltage switchgear problems timely and solve them. The test result shows that high voltage switchgear detecting unit really works.

References

[1] The development of smart switch system for the automation system in power utilities. HASSAN Y B,DINAZZAM M. The Eleventh International Middle East Power System Conference(2006)

[2] Development and implementation of intelligent system for gas insulated switchgear.GUAN Yonggang,HUANG Yulong,LIU Weidong.Power and Energy Engineering Conference(2009)

[3] Continuous monitoring of circuit breakers using vibration analysis. Hidalen,Hans Kristian,Runde, Magne.IEEE Transactions on Power Delivery(2005)

[4] Study on the method of state detection and fault diagnosis for high voltage circuit breakers [D]. Huang Linjie. Beijing Jiaotong University(2007)

[5] Experiences with condition monitoring of HV circuit breakers.Anton Poeltl. IEEE (2001)

[6] Research on the overload protection reliability of moulded case circuit-breakers and its test device[J].Li Kui,Lu Jianguo,Wu Yi,Qin Zhijun,Yao Dongmei. Journal of Zhejiang University SCIENCE A . (2007)